Engineering Maintainability

Engineering Maintainability

Editor

Nitin Jain

Engineering Maintainability
Edited by **Nitin Jain**

Printed in 2017

ISBN: 978-1-68117-362-7

Library of Congress Control Number: 2015941553

© 2016 by
SCITUS Academics LLC,
616, Corporate Way, Suite 2, 4766,
Valley Cottage, NY 10989

www.scitusacademics.com

Notice

Reasonable efforts have been made to publish reliable data and views articulated in the chapters are those of the individual contributors, and not necessarily those of the editors or publishers. Editors or publishers are not responsible for the accuracy of the information in the published chapters or consequences of their use. The publisher believes no responsibility for any damage or grievance to the persons or property arising out of the use of any materials, instructions, methods or thoughts in the book. The editors and the publisher have attempted to trace the copyright holders of all material reproduced in this publication and apologize to copyright holders if permission has not been obtained. If any copyright holder has not been acknowledged, please write to us so we may rectify.

Contents

Preface

Engineering maintenance is an important sector of the economy. This century will usher in a broader need for equipment management—a cradle-to grave strategy to preserve equipment functions, avoid the consequences of failure, and ensure the productive capacity of equipment. This cannot be achieved by simply following the traditional approach to maintenance effectively—human error in maintenance, quality and safety in maintenance, software maintenance, reliability-centered maintenance, maintenance costing, reliability, and maintainability also must be considered. It provides the guidelines and fundamental methods of estimation and calculation needed by maintainability engineers. It also covers the management of maintainability efforts, including issues of organizational structure, cost, and planning processes.

Editor

Band-Gap Engineering of NaNbo$_3$ for Photocatalytic H$_2$ Evolution with Visible Light

Peng Li[1], Hideki Abe[1, 2, 3], and Jinhua Ye[1, 2, 4]

[1]Catalytic Materials Group, Environmental Remediation Materials Unit, National Institute for Materials Science (NIMS), 1-1 Namiki, Tsukuba, Ibaraki 305-0044, Japan

[2]TU-NIMS Joint Research Center, School of Materials Science and Engineering, Tianjin University, 92 Weijin Road, Nankai District, Tianjin 300072, China

[3]PRESTO, Japan Science and Technology Agency (JST), 4-1-8 Honcho Kawaguchi, Saitama 332-0012, Japan

4International Center for Materials Nanoarchitectonics (WPI-MANA), National Institute for Materials Science (NIMS), 1-1 Namiki, Tsukuba, Ibaraki 305-0044, Japan

ABSTRACT

A new visible light response photocatalyst has been developed for H_2 evolution from methanol solution by elemental doping. With lanthanum and cobalt dopants, the photoabsorption edge of $NaNbO_3$ was effectively shifted to the visible light region. It is also found that the photoabsorption edge is effectively controlled by the dopant concentration. Under visible light irradiation, H_2 was successfully generated over the doped $NaNbO_3$ samples and a rate of $12\,\mu mol \cdot h^{-1}$ was achieved over $(LaCo)_{0.03}(NaNb)_{0.97}O_3$. Densityfunctional theory calculations show that Co-induced impurity states are formed in the band gap of $NaNbO_3$ and this is considered to be the origin of visible-light absorption upon doping with La and Co.

INTRODUCTION

Because of the current energy crisis and environmental pollution from the consumption of fossil fuels, new source which can provide a big amount of maintainable energy must be developed in hurry. H_2 is considered as a candidate of the next generation energy source because of its renewable, unlimited, and environmental friendly performances [1, 2]. However, there are still several barriers to realize the practical utilization of H_2 energy, and the produce of H_2 is the most serious one. As the present H_2 is mostly generated from the reformation of fossil fuel, a new method which can produce H_2 with clean energy should be developed [3]. Photocatalysis has been developed as a candidate that can satisfy the demand of supplying H_2 by splitting water with solar energy. In the past decades, a lot of photocatalysts were developed for producing H_2

with high efficiency. But most of the photocatalysts, such as TiO$_2$, SrTiO$_3$, and NaTaO$_3$, have only UV light responsibility, and the low visible light utilization limited the practical use of photocatalysis with solar light [4–6]. To improve the visible light absorption, the common method is doping with cations to adjust the electronic structures of photocatalysts [7]. When the cation dopants replace the positions of lattice cations or occupy the interstices in the crystal lattice, impurity energy levels might be generated within the band gap of the photocatalyst, which can extend the responsive region of photocatalytic reactions into visible light [8, 9].

NaNbO$_3$ is a typical nontoxic and highly stable semiconductor which has abundant applications in photocatalysis. In many reports, NaNbO$_3$ has been demonstrated to be a high efficiency photocatalyst for H$_2$ generation [10–17]. Under the irradiation of UV light, NaNbO$_3$ nanoparticles could reduce H$_2$O to H$_2$ with quite high efficiency with sacrificial agents [12]. Fiber-structured NaNbO$_3$ was also verified to be useful in splitting pure H$_2$O into H$_2$ and O$_2$ [10]. However, almost all the reported NaNbO$_3$ photocatalysts are only sensitive to the UV light. Although iridium doped NaNbO$_3$ was proved to be active in water splitting under visible light irradiation, the efficiency is still low and this method needs precious metal [18]. To achieve visible light photoactivity of NaNbO$_3$ without previous metal dopant is still a big challenge. Cobalt, which is a typical transition element with partially occupied d state, is commonly used as dopant to improve the visible light responsibility of wide band-gap photocatalysts [19–22]. However, simply doping binary oxide with cobalt may increase the defect concentration and negatively affect the photocatalytic performance. Thus, codoping is more popular to balance the charge state and decrease the defects [23, 24]. In this work, we developed a series of NaNbO$_3$ doped with lanthanum and cobalt with H$_2$ evolution activity under visible light irradiation. The further theoretical study indicates that the cobalt dopant creates new states in the band gap of NaNbO$_3$ and provides the visible light absorption.

EXPERIMENTAL SECTION

Material Preparation

The samples were synthesized via a hydrothermal method [12]. In a typical synthesis of $NaNbO_3$, 1.0 g of $(C_2H_5O)_5Nb$ and 0.24 g of C_2H_5ONa were added into 10 mL of 2-methoxyethanol and stirred at room temperature to form a clear colloid. Next, the mixture was stirred for 30 minutes and then heated to 120°C with a rate of 1°C·min^{-1} and maintained at this temperature until a dry gel was obtained. After that, 40 mL of 6 M NaOH solution was added to the powdered dry gel and stirred at room temperature to form a uniform precursor. Then, the mixture was transferred into a 50 mL Teflon sealed autoclave and heated at 180°C for 24 h. Finally, the product was washed with distilled water until pH was lower than 8.0 and the obtained powder was dried at 70°C overnight. To synthesize La, Co codoped $NaNbO_3$, the dopant reagent $La(CH_3COO)_3$, and $Co(CH_3COO)_2$ were added in the first step and all the other procedures were the same.

Sample Characterization

The crystal structure of $NaNbO_3$ powder was determined by an X-ray diffractometer (Rint-2000, Rigaku Co., Japan) with Cu-Ka radiation. The optical absorption spectra were measured with a UV-visible spectrophotometer (UV-2500PC, Shimadzu Co., Japan) using a $BaSO_4$ reference. Scanning electron microscopy images were recorded with a field emission scanning electron microscopy (JSM-6701F, JEOL Co., Japan) operated at 15 kV.

Photocatalytic H$_2$ Evolution

The H_2 evolution experiments were carried out in a gas closed circulation system. In a typical experiment, 0.3 g catalyst was dispersed by a magnetic stirrer in a CH_3OH solution (220 mL distilled water and 50 mL CH_3OH) in a Pyrex cell with a side window. Calculated amount of H_2PtCl_6 solution (0.5 wt%) was added into the solution. The light source used for cocatalyst deposition was a 300 W Xe arc lamp without filter (>300 nm). After the H_2 evolution rate became constant, the system was evacuated and an L-42 cutoff filter was added to the 300 W Xe arc lamp (>420 nm). The H_2 evolution was measured by an in situ gas chromatograph (GC-8A, Shimadzu Co., Japan) with a thermal conductivity detector (TCD).

Theoretical Calculation

The band structures, densities of state (DOS), and partial densities of state (PDOS) of $NaNbO_3$ and codoped $NaNbO_3$ were calculated using the plane-wave density functional theory (DFT) with the CASTEP program package [25]. The doping concentration was set to 3.125% by, respectively, replacing a Na atom and a Nb atom by a La and a Co atom in a 2×2×1 supercell. The electronic state of Co was [Ar] $3d^6$ and high spin. The core electrons were replaced by ultrasoft pseudopotentials with a plane-wave basis cutoff energy of 410 eV, and the interactions of exchange and correlation were treated with Perdew-Burke-Ernzerhof parameterization (PBE) of the generalized gradient approximation (GGA). The FFT grids of basis in all the models were 40 × 40 × 108 and the K-point sets of 3×3×1 were used.

RESULTS AND DISCUSSIONS

The crystallographic structures of all the synthesized $NaNbO_3$ samples were determined by X-ray diffraction (XRD) measurement (as shown in Figure 1(a)). All the observed diffraction peaks in the XRD patterns of $NaNbO_3$ and doped $NaNbO_3$ present good agreement with the reference data from the standard diffraction database (JCPDS-073-0803), showing that every sample was well crystalized in a single phase with the space group of Pbcm, which is the common phase of $NaNbO_3$. However, slight shifts could be found when focusing on the particular diffraction peaks. Figure

1(b) gives the enlarged diffraction peaks with the highest intensity of $NaNbO_3$ and doped $NaNbO_3$. When doping $NaNbO_3$ with La and Co, the diffraction peak shifts to the smaller diffraction angle, suggesting that the unit cell of $NanbO_3$ has a slight expansion. As the radius changes from Na^+ (102 pm) and Nb^{5+} (64 pm) to La^{3+} (103.2 pm) and Co^{3+} (61 pm), such expansion of cell volume is understandable [26]. The XPS measurement (as shown in Figure S1 in Supplementary Material available online at http://dx.doi.org/10.1155/2014/380421) gives obvious evidence that the valance state of Co is +3 as no evident peak of Co^{2+} is observed [27]. The detailed lattice parameters of the as-prepared doped and undoped $NaNbO_3$ samples are shown in Table 1.

Table 1: Crystal structures of the as-prepared doped and undoped NaNbO₃ samples

Materials	Crystal system	Lattice parameters (Å)		
		a	b	c
NaNbO₃	Orthorhombic	5.5028(7)	5.5474(3)	15.4988(6)
$(LaCo)_{0.01}(NaNb)_{0.99}O_3$	Orthorhombic	5.5098(4)	5.5542(2)	15.5047(4)
$(LaCo)_{0.03}(NaNb)_{0.97}O_3$	Orthorhombic	5.5122(5)	5.5650(2)	15.5321(3)
$(LaCo)_{0.05}(NaNb)_{0.95}O_3$	Orthorhombic	5.5128(7)	5.5674(4)	15.5388(3)

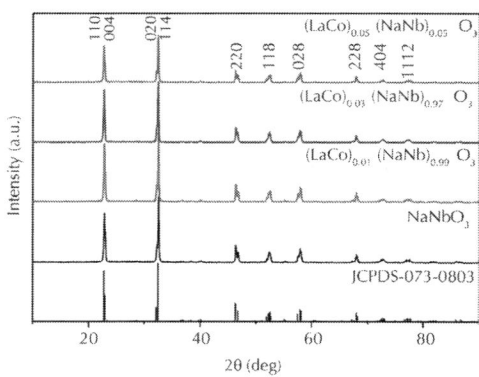

(a)

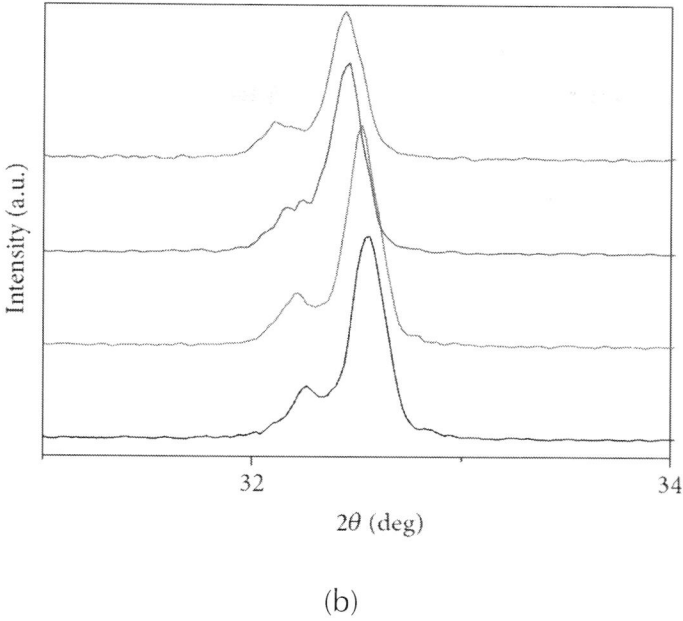

(b)

Figure 1: (a) XRD patterns of the as-prepared NaNbO$_3$ and La, Co codoped NaNbO$_3$ compared with the standard NaNbO$_3$ XRD pattern. (b) The enlarged XRD patterns of the highest diffraction peak of NaNbO$_3$.

Since the morphology is an important factor which can greatly affect the photocatalytic performance, the scanning electron microscope (SEM) was further used to observe the morphology of the as-prepared samples and the SEM images of NaNbO$_3$ and $(LaCo)_{0.05}(NaNb)_{0.95}O_3$ are shown in Figure 2. The NaNbO$_3$ sample is constituted by particles with the cubic morphology, and the cubic particles are generally 300~1000 nm in length. The obtained NanbO$_3$ has the similar morphology as the sample synthesized by hydrothermal reaction in the previous report [12]. Although the crystal structure changes a little after doping with La and Co, the crystal growth process has almost no change. The doped sample has the same morphology as the pure NaNbO$_3$.

(a)

(b)

Figure 2: SEM images of the as-prepared (a) NaNbO$_3$ and (b) La, Co codoped NaNbO$_3$.

UV-visible absorption spectra of NaNbO$_3$ and La, Co codoped NaNbO$_3$ powder samples are shown in Figure3(a). The pure NabO$_3$ sample only has an intense absorption with steep edges in the UV region. Different from the pure NaNbO$_3$, the samples have evident absorptions in the visible light region. The optical band gaps E$_g$ of the as-prepared NaNbO$_3$ samples were determined according to the following equation:

$$(\alpha h v)^n = A\left(h v - E_g\right),$$

(1)

in which α, v, A, and E_g are absorption coefficient, light frequency, proportionality constant, and optical band gap, respectively [28]. The value of index n depends on the property of materials, whereas $n=2$ for the direct transition and $n = 1/2$ for the indirect transition. For NaNbO3, the index n was determined to be 1/2 according to the relationship between lg(αh]) and lg(h] $- E_g$). For La, Co codoped NaNbO$_3$, the indexes were determined to be 2. The different indexes of NaNbO$_3$ and doped NaNbO$_3$ indicate that NaNbO$_3$ is an indirect band-gap semiconductor, while the doped NaNbO$_3$ samples have direct transitions with visible light absorptions. From Figure 3(b), the values of the optical band gaps for NaNbO$_3$, $(LaCo)_{0.01}(NaNb)_{0.99}O_3$, $(LaCo)_{0.03}(NaNb)_{0.97}O_3$, and $(LaCo)_{0.05}(NaNb)_{0.95}O_3$ are determined to be 3.42, 2.74, 2.70, and 2.65 eV, respectively. With the increasing of doping concentration, the optical band gap of NaNbO$_3$ is continuously decreasing.

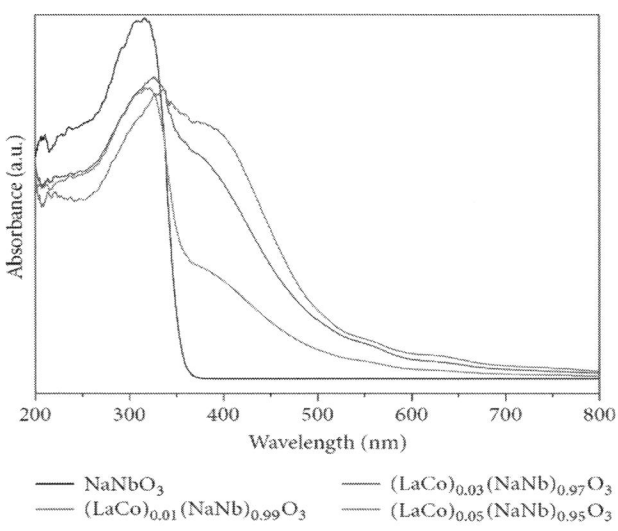

(a)

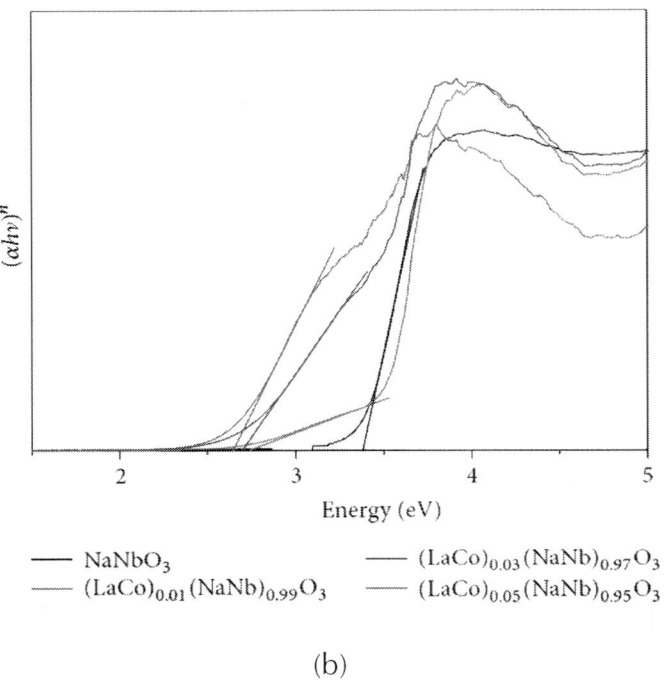

Figure 3: (a) UV-visible absorption spectra of the as-prepared $NaNbO_3$ and La, Co codoped $NaNbO_3$. (b) The corresponding $(\alpha h]) \sim h\nu$ curves of the as-prepared $NaNbO_3$ and La, Co codoped $NaNbO_3$.

The H_2 evolutions from aqueous CH_3OH solution (50 mL CH_3OH + 220 mL H_2O) over $NaNbO_3$ and La, Co codoped $NaNbO_3$ (0.3 g) with 0.5 wt% Pt loading under the irradiation of visible light ($\lambda > 420$ nm) are presented in Figure 4(a). As shown by the UV-visible absorption in the previous part, $NaNbO_3$ has no visible light absorption. Under the irradiation of visible light, there is no H_2 detected during the experiment in 8 hours, while the doped $NaNbO_3$ samples exhibit photoactivities for H_2 evolution in the presence of methanol as sacrificial reagent. H_2 was generated almost linearly over all the doped samples in 8 hours. As plotted in Figure 4(b), the H_2 evolution rates are significantly different: $(LaCo)_{0.03}(NaNb)_{0.97}O_3$ > $(LaCo)_{0.05}(NaNb)_{0.95}O_3$ > $(LaCo)_{0.01}(NaNb)_{0.99}O_3$. Over the best catalyst $(LaCo)_{0.03}(NaNb)_{0.97}O_3$, 11.9 μmol H_2 could be produced every hour.

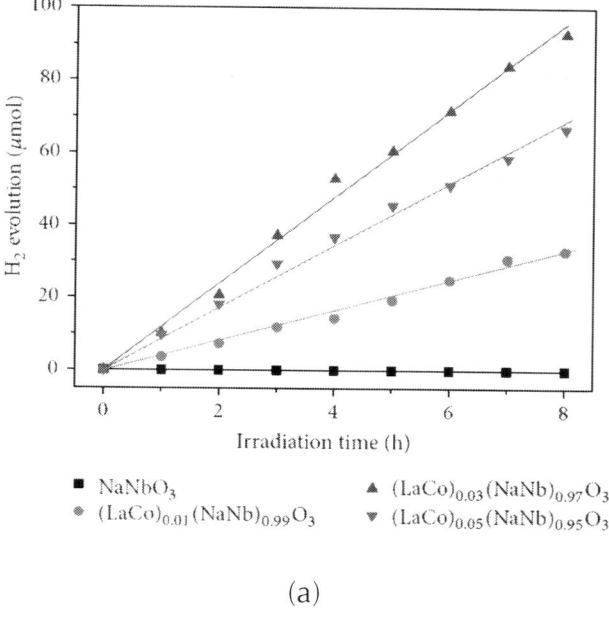

(a)

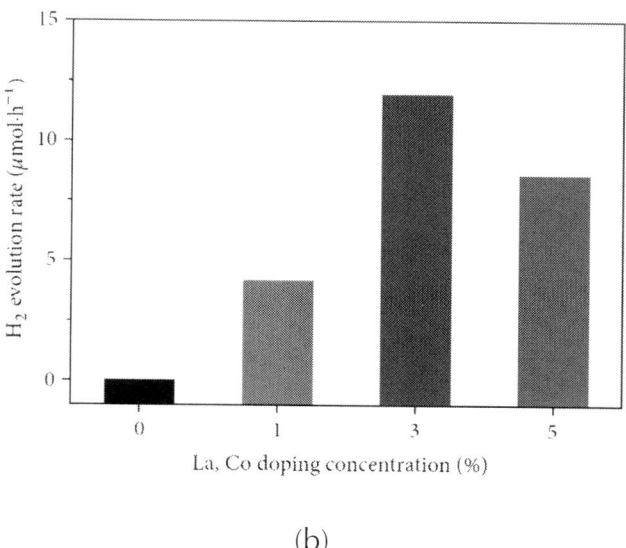

(b)

Figure 4: (a) Photocatalytic H$_2$ evolutions from the aqueous methanol solution over the as-prepared NaNbO$_3$ and La, Co codoped NaNbO$_3$ with 0.5 wt% Pt loading under the irradiation of visible light ($\lambda > 420$ nm).

(b) The comparison of average photocatalytic H_2 evolution rates from the aqueous methanol solution over $NaNbO_3$ and La, Co codoped $NaNbO_3$ with 0.5 wt% Pt loading under the irradiation of visible light ($\lambda > 420$ nm).

To understand the mechanism of visible light photocatalytic activity of La, Co codoped $NaNbO_3$, theoretical calculation based on density functional theory (DFT) was carried out. The density of states (DOS) in Figure 5 indicates that the undoped NaNbO3 has simple valence band maxima (VBM) and conduction band minima (CBM). Its VBM and CBM are mainly composited by O p states and Nb d states. Under light irradiation, the electrons are excited from O p orbitals to Nb d orbitals and the holes are left in O p orbitals. Then, the photogenerated electrons and holes migrate to the surface and react with water and sacrificial reagent, respectively. With La and Co doping, significant changes could be found with VBM and CBM. Two dopant states are observed between the original VBM and CBM, which narrow the band gap of doped $NaNbO_3$ and induce the visible light absorption and visible light response H_2 evolution activity. However, these two states are hybrid by Co d states and O p states and Co d states have larger combination ratio. Thus, the improved visible light absorption is mostly caused by the d-d transition of Co. Since the electrons excited from d states to d states have a high backward transition rate, the photogenerated electrons could hardly migrate to the surface and perform photocatalytic reactions. This is the reason why the photoactivity of La, Co codoped $NaNbO_3$ under visible light is not as high as pure NaNbO3 under UV light. The general mechanism of the visible light activity over La, Co codoped $NaNbO_3$ could be concluded in Figure 6. The doping with Co element creates new occupied and unoccupied energy levels in the band gap of $NaNbO_3$. The transition between the new CBM and VBM could absorb visible and make the visible light photocatalytic reaction possible.

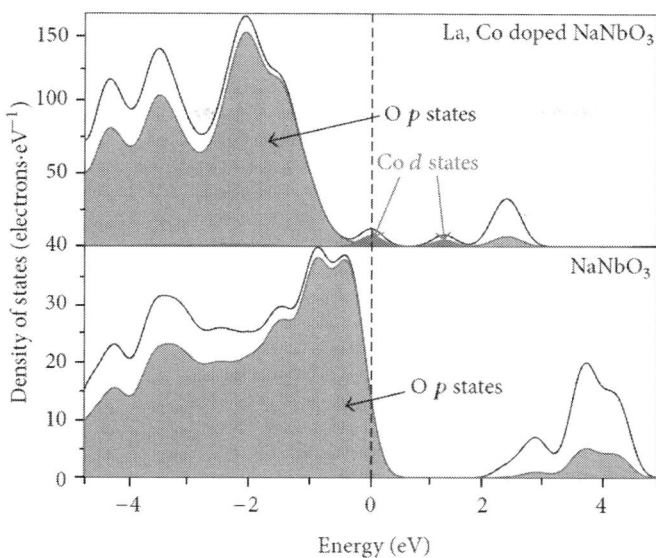

Figure 5: The calculated density of states and partial density of states of NaNbO₃ and La, Co codoped NaNbO₃.

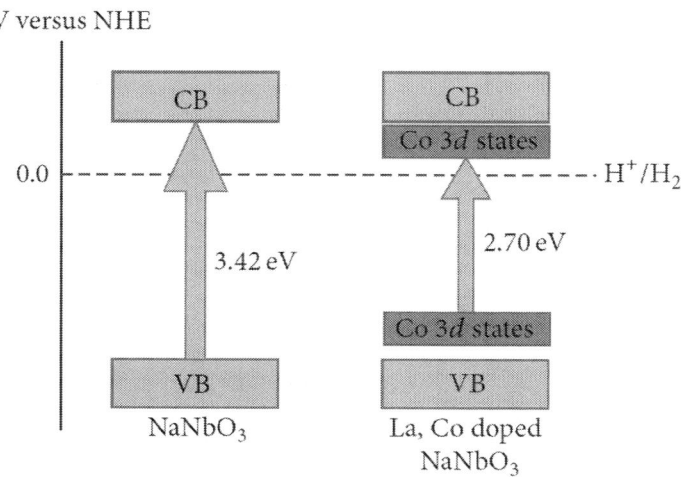

Figure 6: The schematic band structures of NaNbO₃ and La, Co codoped NaNbO₃.

CONCLUSIONS

In conclusion, La, Co codoped $NaNbO_3$ were synthesized to realize the visible light response photocatalytic H_2 evolution. The doped $NaNbO_3$ samples showed narrower optical band gaps (2.65, 2.70, and 2.74 eV for $(LaCo)_{0.05}(NaNb)_{0.95}O_3$, $(LaCo)_{0.03}(NaNb)_{0.97}O_3$, and $(LaCo)_{0.01}(NaNb)_{0.99}O_3$, resp.) than the pure $NaNbO_3$ (3.42 eV). In photocatalytic H_2 evolution experiments, the doped $NaNbO_3$ samples showed activity under the visible light irradiation, while the undoped $NaNbO_3$ was not active. According to the theoretical calculation, the visible light activity of La, Co codoped $NaNbO_3$ could be attributed to the new impurity electronic states of Co dopant. Therefore, this work presented a new material for visible light photocatalytic H_2 evolution.

ACKNOWLEDGMENTS

The authors thank Professor Naoto Umezawa for the result discussion and Dr. Akihiro Tanaka and Dr. Hideo Iwai of Materials Analysis Station of NIMS for the XPS measurement and analysis. This work was supported by Japan Science and Technology Agency (JST) and Precursory Research for Embryonic Science and Technology (PRESTO) program.

REFERENCES

1. A. J. Bard and M. A. Fox, "Artificial photosynthesis: solar splitting of water to hydrogen and oxygen,"Accounts of Chemical Research, vol. 28, no. 3, pp. 141–145, 1995

2. T. J. Meyer, "Chemical approaches to artificial photosynthesis," Accounts of Chemical Research, vol. 22, pp. 163–170, 1989

3. M. D. Hernández-Alonso, F. Fresno, S. Suárez, and J. M. Coronado, "Development of alternative photocatalysts to TiO$_2$:

challenges and opportunities," Energy and Environmental Science, vol. 2, no. 12, pp. 1231–1257, 2009

4. Y. Yoshida, M. Matsuoka, S. C. Moon, H. Mametsuka, E. Suzuki, and M. Anpo, "Photocatalytic decomposition of liquid-water on the Pt-loaded TiO_2 catalysts: effects of the oxidation states of Pt-species on the photocatalytic reactivity and the rate of the back reaction," Research on Chemical Intermediates, vol. 26, no. 6, pp. 567–574, 2000

5. H. Kato and A. Kudo, "New tantalate photocatalysts for water decomposition into H_2 and O_2,"Chemical Physics Letters, vol. 295, no. 5-6, pp. 487–492, 1998

6. K. Domen, S. Naito, M. Soma, T. Onishi, and K. Tamaru, "Photocatalytic decomposition of water vapour on an NiO–$SrTiO_3$ catalyst," Journal of the Chemical Society, Chemical Communications, no. 12, pp. 543–544, 1980

7. H. Tong, S. Ouyang, Y. Bi, N. Umezawa, M. Oshikiri, and J. Ye, "Nano-photocatalytic materials: possibilities and challenges," Advanced Materials, vol. 24, no. 2, pp. 229–251, 2012

8. J. Y. Cao, Y. J. Zhang, H. Tong, P. Li, T. Kako, and J. H. Ye, "Selective local nitrogen doping in a TiO_2electrode for enhancing photoelectrochemical water splitting," Chemical Communications, vol. 48, pp. 8649–8651, 2012

9. J. W. Shi, J. H. Ye, L. J. Ma, S. X. Ouyang, D. W. Jing, and L. J. Guo, "Site-selected doping of upconversion luminescent Er^{3+} into $SrTiO_3$ for visible-light-driven photocatalytic H_2 or O_2 evolution,"Chemistry, vol. 18, no. 24, pp. 7543–7551, 2012

10. H. F. Shi, X. K. Li, D. F. Wang, Y. P. Yuan, Z. G. Zou, and J. H. Ye, "$NaNbO_3$ nanostructures: facile synthesis, characterization, and their photocatalytic properties," Catalysis Letters, vol. 132, pp. 205–212, 2009

11. P. Li, S. Ouyang, G. Xi, T. Kako, and J. Ye, "The effects of crystal structure and electronic structure on photocatalytic H_2 evolution and CO_2 reduction over two phases of perovskite-structured $NaNbO_3$," The Journal of Physical Chemistry C, vol. 116, no. 14, pp. 7621–7628, 2012

12. G. Li, T. Kako, D. Wang, Z. Zou, and J. Ye, "Synthesis and enhanced photocatalytic activity of NaNbO3 prepared by hydrothermal and polymerized complex methods," Journal of Physics and Chemistry of Solids, vol. 69, no. 10, pp. 2487–2491, 2008

13. P. Li, H. Xu, L. Liu et al., "Constructing cubic-orthorhombic surface-phase junctions of $NaNbO_3$ towards significant enhancement of CO_2 photoreduction," Journal of Materials Chemistry A, vol. 2, no. 16, pp. 5606–5609, 2014

14. N. Chen, G. Li, and W. Zhang, "Effect of synthesis atmosphere on photocatalytic hydrogen production of $NaNbO_3$," Physica B, vol. 447, pp. 12–14, 2014

15. G. Li, W. Wang, N. Yang, and W. F. Zhang, "Composition dependence of $AgSbO_3/NaNbO_3$ composite on surface photovoltaic and visible-light photocatalytic properties," Applied Physics A: Materials Science & Processing, vol. 103, pp. 251–256, 2011

16. G. Li, Z. Yi, Y. Bai, W. Zhang, and H. Zhang, "Anisotropy in photocatalytic oxidization activity of $NaNbO_3$ photocatalyst," Dalton Transactions, vol. 41, no. 34, pp. 10194–10198, 2012

17. X. Li, G. Li, S. Wu, X. Chen, and W. Zhang, "Preparation and photocatalytic properties of platelike $NaNbO_3$ based photocatalysts," Journal of Physics and Chemistry of Solids, vol. 75, pp. 491–494, 2014.

18. A. Iwase, K. Saito, and A. Kudo, "Sensitization of $NaMO_3$ (M: Nb and Ta) photocatalysts with wide band gaps to visible light by Ir doping," Bulletin of the Chemical Society of Japan, vol. 82, pp. 514–518, 2009

19. J. Choi, H. Park, and M. R. Hoffmann, "Effects of single metal-ion doping on the visible-light photoreactivity of TiO_2," The Journal of Physical Chemistry C, vol. 114, no. 2, pp. 783–792, 2010

20. D. Dvoranova, V. Brezova, M. Mazur, and M. A. Malati, "Investigations of metal-doped titanium dioxide

photocatalysts," Applied Catalysis B, vol. 37, no. 2, pp. 91–105, 2002

21. M. Iwasaki, M. Hara, H. Kawada, H. Tada, and S. Ito, "Cobalt ion-doped TiO$_2$ photocatalyst response to visible light," Journal of Colloid and Interface Scienc, vol. 224, pp. 202–204, 2000

22. B. Zhou, X. Zhao, H. Liu, J. Qu, and C. P. Huang, "Visible-light sensitive cobalt-doped BiVO$_4$ (Co-BiVO$_4$) photocatalytic composites for the degradation of methylene blue dye in dilute aqueous solutions," Applied Catalysis B, vol. 99, pp. 214–221, 2010

23. Z. G. Yi and J. H. Ye, "Band gap tuning of Na$_{1-x}$La$_x$Ta$_{1-x}$Co$_x$O$_3$ solid solutions for visible light photocatalysis," Applied Physics Letters, vol. 91, Article ID 254108, 2007

24. Z. G. Yi and J. H. Ye, "Band gap tuning of Na$_{1-x}$La$_x$Ta$_{1-x}$Cr$_x$O$_3$ for H$_2$ generation from water under visible light irradiation," Journal of Applied Physics, vol. 106, Article ID 074910, 2009

25. M. D. Segall, P. J. D. Lindan, M. J. Probert et al., "First-principles simulation: ideas, illustrations and the CASTEP code," Journal of Physics Condensed Matter, vol. 14, no. 11, pp. 2717–2744, 2002

26. R. Shannon, "Revised effective ionic radii and systematic studies of interatomic distances in halides and chalcogenides," Acta Crystallographica A, vol. 32, pp. 751–767, 1976

27. M. C. Biesinger, B. P. Payne, A. P. Grosvenor, L. W. M. Lau, A. R. Gerson, and R. S. C. Smart, "Resolving surface chemical states in XPS analysis of first row transition metals, oxides and hydroxides: Cr, Mn, Fe, Co and Ni," Applied Surface Science, vol. 257, no. 7, pp. 2717–2730, 2011

28. M. A. Butler, "Photoelectrolysis and physical properties of the semiconducting electrode WO$_2$," Journal of Applied Physics, vol. 48, pp. 1914–1920, 1977

Design and Performance Evaluation of a Sustained Load Dual Grip Creep Testing Machine

Kenneth Kanayo Alaneme[1, 2], Bethel Jeremiah Bamike[1], and Godwin Omlenyi[1]

[1]Department of Metallurgical and Materials Engineering, Federal University of Technology, Akure, Nigeria

[2]Department of Mining and Metallurgical Engineering, University of Namibia, Ongwediva Engineering Campus, Ongwediva, Namibia

ABSTRACT

The design and performance evaluation of a sustained load creep testing machine was undertaken in this research. The design was motivated by the need to make locally available, a cost effective, technically efficient, and easily operated creep testing facility; for creep behaviour studies of materials. Design drawings and purchase of materials and components for the design were undertaken after thorough evaluation of the following design and materials selection criteria: design principle and theory, local availability of raw materials and components required for the design, material properties, cost of materials and design, ease of utilization and maintenance, and basis of testing and data capture. The machine casing and frame, heating chamber (consisting of the furnace and a dual specimen mounting stage), load lever and hanger system, and the electro-technical components; were fabricated and coupled following the produced design specifications. The machine was tested and its performance was assessed using its heating efficiency, repeatability and reproducibity of experimental test results, maintainability and cost-effectiveness as criteria. It was observed from repeat tests that the machine has the capacity of generating reliable data for computing creep strain-time results. The efficiency and temperature regulating capacity of the heat- ing unit of the machine were also observed to be very satisfactory. The cost of the design was about 112,000 Naira ($700.00) which is cheaper in comparison to similar commercial creep testing machines from abroad. The machine was also found not to pose maintenance or repairs challenges.

INTRODUCTION

Creep has been acknowledged to be the most active failure mechanism of engineering materials under stress at elevated temperature conditions [1]. The considerable material flow which occurs over a period of time in creep situations can have grave material performance and service life implications [2]. Thus there is a lot of interest in understanding the creep behaviour of materials for high temperature applications. From a mechanical behaviour of materials perspective, creep mechanism is influenced by the increased atom mobility, vacancy density, and ease of dislocation glide or climb at elevated temperature [3] [4]. These phenomena often combine to facilitate permanent deformation which results in material rupture in severe creep cases [4].

Creep failure of components/parts in many industries such as metallurgical processing, power generation, petrochemical, spacecraft, and nuclear plants has been well reported in literature [5] [6] . In architectural and building designs, a good number of polymer/polymer composite materials are currently used as structural and semi-structural components. Due to exposure to intermittent solar radiation, the creep behaviour of these polymer based materials has also come under scrutiny [7]. It is thus imperative in materials design for high temperature applications, to account for creep behaviour to safeguard against likely failure short of projected design life time.

The basis for creep testing of materials including the specifications of facilities, specimens, and testing procedures has been assessed over the years following recommendations in ASTM standard codes [8] [9]. Creep tests are essentially used to establish a suitable design stress for components for a specified time and at given temperature or for specifying a maximum permissible strain for maintaining functionality of components during its service life [9].

Majorly due to high cost of purchase and maintenance of test facilities and accessories for creep testing, there exists a paucity of researches on creep studies of the myriad of materials developed within Africa by most African material engineers. Efforts to evaluate the creep properties of these indigenously developed materials outside the shores of Africa, do not always attract much interest from potential host organizations. There is also the need to conserve foreign exchange and work towards technological self-reliance. This research work is an effort to address these problems through the design of a creep testing machine using locally sourced raw materials and components. The design concept for this creep testing machine is based on the application of a constant load from a cantilever type lever on specimens mounted in an enclosed heating chamber (furnace). Some of the potential benefits of the design are low cost, maintainability, accessibility, and adaptability for research and experimental demonstrations.

MATERIALS AND METHOD

Materials

The materials used for the design of the creep testing machine are: mild steel sheets, square steel pipes, porous refractory bricks, fire clay, Kaolin, Nichrome heating elements, K-type thermocouple, light Indicators, temperature controller, dial gauge, timer, round and flat Chuck Grips, and stainless steel rods, copper wire and plugs, dead weights, and weight hangers.

Design Considerations

At the conceptualization stage of producing a design for the creep testing machine, the factors considered were: the design principle and theory, local availability of potential materials required for the fabrication, material properties, cost of materials, simplicity of the design, ease of utilization and maintenance of the machine, and basis of generating required data. These factors were instructive in the development of the design of the creep machine (Figure 1). The main parts of the cantilever type sustained load creep testing machine are: the machine casing and frames, the heating unit (which consists of a furnace with an inbuilt specimen clamping/mounting system), the loading frame which consists of the loading beam and the load hanger (which serves as the anchor from which the applied load is transmitted to the test specimen), dial gauge (to measure the strain generated by the specimen during testing), timer (which is utilized to record the time to failure or attaining a maximum permissible strain on the specimen), the electro-technical devices (which consists of the temperature controller, thermocouple and the light indicators) to set and maintain the temperature in the heating unit of the machine, and dead weights utilized to apply predetermined loads on the test specimens. These components/parts of the creep machine are indicated in the design (Figure 1(b)) with the aid of the legends identified in Table 1.

Design and Materials Selection Criteria

Machine Casing and Frame

Mild steel sheets (2 mm thick) were selected for the design of the machine casings. The choice of mild steel was based on low cost of purchase, good strength, weldability, excellent formability, and availability. The steel casing houses all the components of the creep testing machine including: the heating chamber made up of the furnace and specimen mounting system, the electro-technical devices (temperature controller, thermocouple, and light indicators), and the loading system and platform. Steel square pipes (equally 2 mm thick), was selected for the construction of the support frames of the machine. The steel frames serves to support the entire weight of the creep testing system. It also gives rigidity and balance to the entire creep testing unit.

Furnace Design

Refractory bricks, clay and kaolin were selected as lining materials for the design of the heating unit of the machine. The choice of the materials was influenced by low cost considerations, local availability, high refractory properties, and low thermal conductivity.

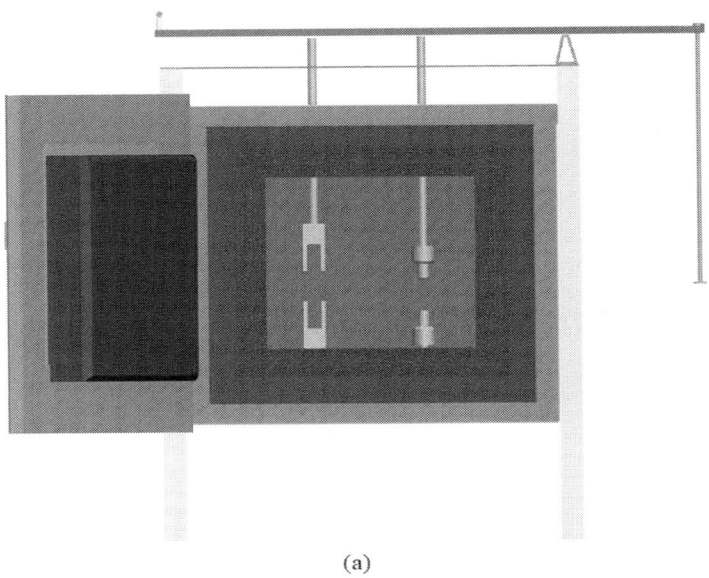

(a)

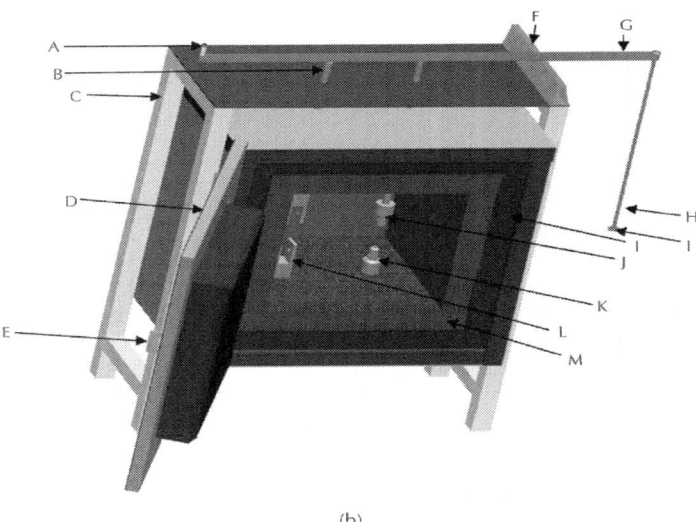

(b)

Figure 1: Showing modified: (a) Front view design of the creep testing machine; and (b) 3D full view design of the creep testing machine.

Table 1: Legend of the 3D design of the creep testing machine presented in Figure 1(b)

Label	Name
A	Dial Gauge
B	Grip Extension
C	Metal Frame
D	Furnace Door
E	Door Handle
F	Pivot
G	Lever
H	Load Pipe
I	Load Plate
J	Upper Circular Grip
K	Lower Circular Grip
L	Flat Grip
M	Refractory

Clamps and Clamping System

The clamping system of the machine consists of chuck grips needed to hold the test samples firmly in place during the creep experiments. Two different sample mounting platforms were designed having different grip configurations (one for cylindrical specimens while the other serves for flat specimens). This is to make it possible for different specimen configurations (flat and round) to be conveniently tested using either of the sample grip mounting platforms. The base of the mounting platforms is firmly bolted at the bottom of the casing of the heating chamber while the top end is flexibly connected to the load cantilever beam which is linked to the loading platform of the machine. The flexible connection allows for load applied through the load lever to be transmitted and sustained by the test sample only [10] . The chucks and connecting rods of both gripping systems (flat and round) were properly aligned and installed on the mounting stage of the machine. Chucks and pull rods made of high carbon low alloy steel (in the quenched and tempered condition) were selected for the design of the sample mounting system. Since the primary intent was basically to use the creep machine for low melting point light weight metallic materials and polymers; the creep resistance, thermomechanical fatigue resistance, oxidation resistance, high melting point, and relative low cost of the high carbon low alloy steel influenced its selection ahead of other competing materials.

Cantilever Beam Loading System

The loading system of the machine consists of a cantilever type load beam, a load hanger, and variable dead weights. Quenched and tempered high carbon low alloy steel materials were selected for the design of the load beam and load hanger. This is to ensure that the load frame possesses high strength, elastic modulus and rigidity; and is able to sustain the applied load over a long period of time without undergoing plastic deformation (bending) or fracture while in use [11] . This safeguards against potential erroneous strain capture by the dial gauge from the deformation of the load beam or load hanger as against that of the sample. The dimensions of the loading frame were carefully selected so that the weight of the loading frame does not add significantly to the weight acting on the test specimen as this could affect the calculated stresses acting on the specimen.

Strain Measuring Device

A dial gauge with measurement sensitivity of 0.001 mm was selected as strain recording device. It is connected at the opposite arm of the loading point of the cantilever beam so as to capture changes in the strain of the test sample. When load is applied to the specimen, strain occurs and relative movement between the gripping points is transmitted through the load beam to the dial gauge. A portable precision hour-minute-second timer was selected to monitor the time dependence of the strains developed on the specimens.

Electro-Technical System

The electro-technical system consists of the temperature controller, the heating elements, the thermocouple, and the light indicators. A Nichrome coil was selected as heating element for the furnace. Nichrome was selected as resistance heating element for the furnace because it efficiently converts electricity into heat. It possesses good ductility and formability-requirements needed to form it into coils of any shape and size. It also has a relatively high melting point, can maintain its original shape and dimensional integrity even after several heating cycles, and has a high oxidation resistance [12] . The temperature controller selected as part of the electro-technical devices has the capacity of sensing fractions of temperature which helps in improving the precision and accuracy of the readings sensed by the thermocouple during the operation of the furnace. The connection of the electro- technical devices (temperature controller, thermocouple and the light indicators) help to set, regulate and monitor the temperature in the heating unit of the machine to ensure it is within acceptable limits of the set value (target temperature).

Fabrication Procedure

The casing for the heating chamber was first positioned and lined with refractory bricks and then plastered using a mixture of kaolin, clay, and water which served as binder in accordance with Alaneme et al. [13] . Grooves were created around the lined refractory bricks for the housing of the heating elements. The heating elements were passed through the grooves to allow for efficient and even heat generation. The gripping devices for the mounting of the specimens for testing were then positioned within the chamber with the bottom end bolted firmly at the base of the heating chamber (fixed end) and the top portion of the gripping system connected to the movable load lever system with the help of a hinge [14] . The furnace was then covered at the top with bricks and a metallic sheet and the moveable part of the grips (upper grips) were held in place. The electrical connection to the heating element was done and linked to the electro-technical devices which were placed in a steel box casing by the side of the heating chamber. The assembly of the electro-technical devices in its housing required the connection of the thermocouple through the thermocouple lead to the temperature controller. The heating chamber (furnace) is powered through an industrial switch linked to an AC power source. The progression in heating measured by temperature is monitored with the help of the LED light indicator and temperature controller display. On completion of the assembly of the various components of the machine, it was cleaned using emery papers to obtain a smooth finish and then sprayed to improve the finishing. The interior view of the heating chamber and the external view of the fabricated machine are presented in Figure 2.

(a)

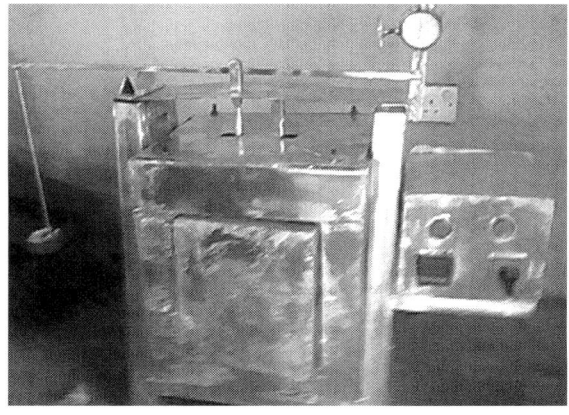

(b)

Figure 2: Showing (a) completed interior view of the creep testing machine and (b) complete full exterior view of the creep testing machine.

Testing of the Machine

A number of polymer based materials, polypropylene/polypropylene based composites and polytetra fluoroethylene (pictured in Figure 3) were selected as test material for assessing the operation of the machine. The flat test samples were machined having gauge length of 40 mm and 2 mm thickness, and tested following specifications of the ASTM D2990-09 standard [8] . The machined test sample was mounted on the chucks of the machine, before the furnace is switched on and the desired temperature set on the temperature controller. A known weight is applied on the specimen through the load hanger and to ensure accuracy of the readings from the test, the dial gauge is set at zero for accurate extension capture. The timer is equally switched on to assist the recording of extension values at specified time intervals. The specimen strain at the specified time intervals is determined by dividing the sample extension by the original gauge length of the sample. Several repeat tests were performed to ascertain the consistency in results obtained under same test conditions.

RESULTS AND DISCUSSION

Machine Performance

The consistency of repeat test results obtained from the testing of samples showed that the machine can be used to generate reliable strain-exposure time data needed for studying the creep behaviour of materials. Representative results from creep testing of polyfluoroethylene (Figure 4) is observed to follow the characteristic creep deformation sequence and the creep strain rate increases with applied stress at a constant temperature (100°C) [2]. The creep behaviour observed from the test material (polyfluoroethylene) is comparable with creep characteristics of similar materials reported in literature [15] . It was also observed that the furnace has a high heating rate and is able to maintain a set temperature value at an accuracy of ±2°C. The furnace had good heat retaining capacity an indication that the refractory selected has good refractory properties and the lining of the furnace was properly done leaving little room for heat losses [16].

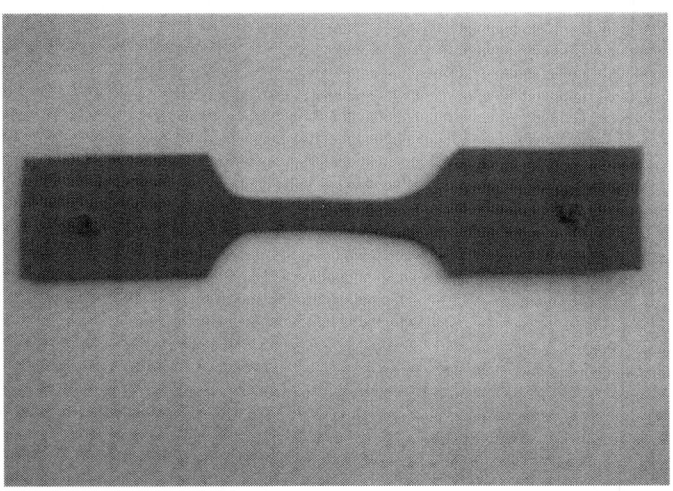

(a)

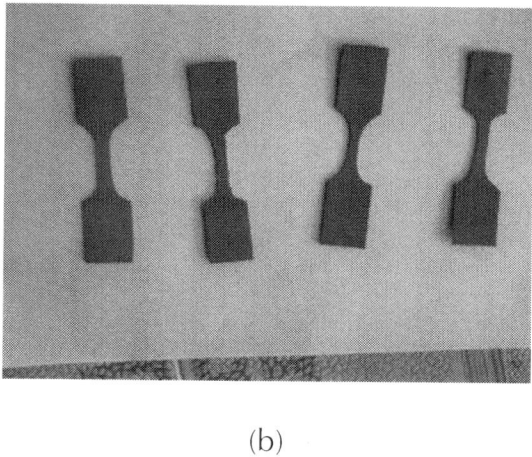

(b)

Figure 3: Representative test samples of polytetrafluoroethylene (Teflon) used for Creep Testing.

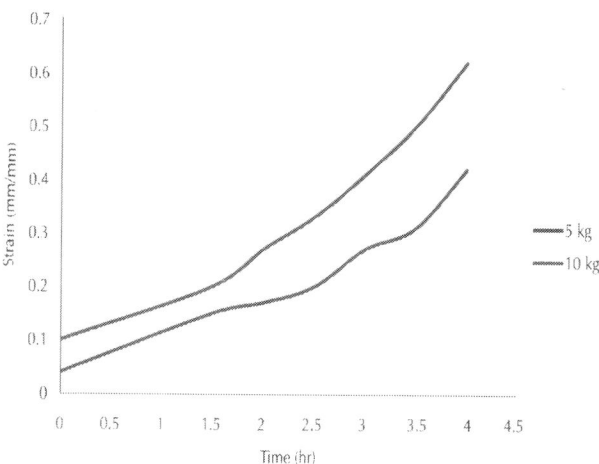

Figure 4: Representative creep strain versus time plots for Polytetrafluoroethylene (Teflon) at different applied load at constant temperature of 100°C.

The machine function was optimized by careful application of some operational strategies especially with the heating unit which is the automated part of the machine. The thermocouple tip is positioned close to the position of the gripping system where the specimens are mounted to ensure that the temperature of the specimen is at the set point temperature value and not just the temperature of the furnace environment that is sensed. Regular calibration of the temperature controller using an external probe is performed to ensure reliability of the temperature readings obtained from the furnace. When testing is to be performed thorough care is taken to ensure that the specimens are tightly clamped in the chuck to safeguard against removal of specimen when the machine is in operation. It was also ensured that the whole machine set up was securely fasted to the machine frame to ensure safety of operator and machine during testing. The mode of operation of the machine can be easily comprehended and does not require complicated basis for data recording. In the case of machine malfunction, the design was made such that all parts can be easily detached and repaired. The replacement of any of the machine parts and fabrication materials when required can be done easily as all parts used in the design of the machine are relatively cheap and can be sourced locally.

Cost Analysis

The entire materials and components used for the design of the creep testing machine are presented in Table 2. The materials and equipment used in the design are locally sourced, and the overall cost of designing the machine is approximately 112,000 Naira ($700.00). The machine is obviously cheaper in comparison to similar commercial brands of creep testing machines designed abroad.

CONCLUSIONS

The design and performance evaluation of a creep testing machine was investigated. The design was motivated by the need to make locally available, a cost effective, technically efficient, and easily operated creep testing facility; for creep behaviour studies of materials. On testing and assessing the performance of the machine, it was observed from repeat tests that the machine has the capacity of generating reliable data for computing creep strain-time results. The efficiency and temperature regulating capacity of the heating unit of the machine were also observed to be very satisfactory. The cost of the design was about 112,000 Naira ($700.00) which is cheaper in comparison to similar commercial creep testing machines from abroad. The machine was also found not to pose maintenance or repairs challenges.

Table 2: Bill for engineering management and evaluation

S/N	Material	Specification	Quantity	Unit Cost (N)	Amount (N)
1	Porous Bricks	Insulating	20	750	15,000
2	Heating Element (Nichrome)	220 - 240 V, 2 KW	2		33,000
3	Jaw Chuck		2	1,500	3000
3	Bits (15 mm Diameter)		1	1,200	1200
4	Contactor	40 Amp	1	2,500	2500
5	Thermometer	1200°C	1	8,000	8000
6	Thermocouple	K-type	1	2,000	2000
7	Sheet Metal	2 mm thick	0.5	3,000	3000
8	Square Pipe	2 mm thick	2	2,000	4000
9	Copper Cable	OFHC		700	1200
10	Dial Gauge	Calibrated	1	5,000	5000
11	Painting	Oil paint	2 litres		1000
12	Transportation/ Logistics				12,000
13	Labour				20,000
14	Total				110,900

REFERENCES

1. Rosler, J., Harders, H. and Baker, M. (2007) Mechanical Behaviour of Engineering Materials—Metals, Ceramics, Polymers, and Composites. Springer, Germany, 333-375.

2. Soboyejo, W. (2002) Mechanical Property of Materials. Princeton University, Princeton, 468-480.

3. Naumenko, K. and Altenbach, H. (2007) Modeling of Creep for Structural Analysis. Springer, New York. http://dx.doi.org/10.1007/978-3-540-70839-1

4. Dieter, G.E. (1986) Mechanical Metallurgy. 3rd Edition, McGraw-Hill, New York.

5. Ravi, S., Laha, K., Sakthy, S., Mathew, M.D. and Jayakumar, T. (2014) Design of Creep Machine and Creep Specimen Chamber for Carrying out Creep Tests in Flowing Liquid Sodium. Nuclear Engineering and Design, 267, 1-9.http://dx.doi.org/10.1016/j.nucengdes.2013.10.020

6. Evans, R.W. and Wilshire, B. (1993) Introduction to Creep. The Institute of Materials, London, 1-75.

7. Kawai, M. (2001) Off-Axis Creep Behavior of Unidirectional Polymer Matrix Composites at High Temperature. Solid Mechanics and its Applications, 86, 469-478.

8. ASTM D2990-09 (2009) Standard test Methods for Tensile, Compressive, and Flexural Creep and Creep Rupture of Plastics. Annual Book of ASTM Standards, ASTM International, West Conshohocken.

9. ASTM E139-11 (2011) Standard Method for Conducting Creep, Creep-Rupture, and Stress-Rupture Tests of Metallic Materials, Annual Book of ASTM Standards, ASTM International, West Conshohocken.

10. Alaneme, K.K. (2011) Design of a Cantilever-Type Rotating Bending Fatigue Testing Machine. Journal of Minerals & Materials Characterization & Engineering, 10, 1027-1039.

11. Alaneme, K.K. (2011) Development of a Cantilever Beam-Sustained Load Stress Corrosion Testing Rig. Journal of Metallurgy and Materials Engineering, 6, 22-26.

12. Srivasta, S. (2014) Properties of Nichrome Wire.http://www.buzzle.com/articles/properties-of-nichrome-wire.html

13. Alaneme, K.K., Olanrewaju, S.O. and Bodunrin, M.O. (2011) Development and Performance Evaluation of a Salt Bath Furnace. International Journal of Mechanical and Materials Engineering, 6, 67-74.

14. Momoh, J.J., Shuaib-Babata, L.Y. and Adelegan, G.O. (2010) Modification and Performance Evaluation of a Low Cost Electro-Mechanically Operated Creep Testing Machine. Leonardo Journal of Science, 16, 83-94.

15. Sakai, T. and Somiya, S. (2011) Analysis of Creep Behavior in Thermoplastics Based on Visco-Elastic Theory. Mechanics of Time-Dependent Materials, 15, 293-308.http://dx.doi.org/10.1007/s11043-011-9136-y

16. Alaneme, K.K. and Olanrewaju, S.O. (2010) Design of a Diesel Fired Heat-Treatment Furnace. Journal of Minerals and Materials Characterization and Engineering, 9, 581-591.

Chapter 3

Radiotracer Technology in Mixing Processes for Industrial Applications

N. Othman[1] and S. K. Kamarudin[2]

[1]Malaysian Nuclear Agency, Bangi, Kajang, 43000 Selangor, Malaysia

[2]Department of Chemical and Process Engineering, Universiti Kebangsaan Malaysia (UKM), Bangi, 43600 Selangor, Malaysia

ABSTRACT

Many problems associated with the mixing process remain unsolved and result in poor mixing performance. The residence time distribution (RTD) and the mixing time are the most important parameters that determine the homogenisation that is achieved in the mixing vessel and are discussed in detail in this paper. In addition, this paper reviews the current problems associated with conventional tracers, mathematical models, and computational fluid dynamics simulations involved in radiotracer experiments and hybrid of radiotracer.

INTRODUCTION

Radiotracers are widely used for the measurement of the flow rate of liquids, gases, and solids in many industrial systems. Thus, in the enhancement of production efficiency and process optimisation, radioisotope-based technology continues to play a rapidly growing role in various industries, such as petrochemicals, oil, and gas, as well as wastewater treatment plants [1–3]. An investigation of many major industrial applications, including fluidised beds, sugar crystallisers, trickle bed reactors, cement rotary kilns, wastewater treatment units, and interwell communications in oil fields, can be performed by injecting a radiotracer at the inlet of the system and monitoring it at the outlet. The data output can be treated and analysed to investigate the behaviour of the system. Figure 1 shows the fundamentals of a tracer experimental setup as described by Furman et al. [4] in which at least one detector is needed at the inlet to detect the presence of the radioactive source prior to the process investigation and at least one detector is needed at the outlet to detect the radioactive source during the study. The peaks signify the detection of the emitted gamma ray from the radiotracer. Normally, the detector used is a scintillation sodium iodide (Tl) detector.

Radiotracer technology also assists industries in satisfying the critical need for production efficiency through the identification of process malfunctions and anomalies, as well as mechanical damage in the plant. Although radioisotopes have been used to solve a number of industry problems for over 50 years, research and development of the technology continue unchallenged. The greatest benefit of radiotracer technology over the conventional methods is that the investigation can be carried out on-stream and without disrupting the operating process of the plant.

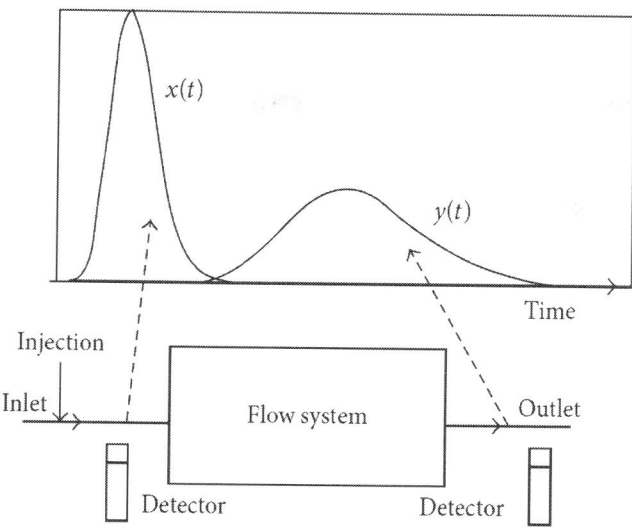

Figure 1: The principle of radiotracer experiment by Furman et al. 2003 [4].

Hence, any expensive downtime is avoided and the convenience of direct measurement results in substantial economic benefits and investigating costs. Nevertheless, although the technology is applicable across a broad industrial spectrum, Pant et al. [5] and Pant et al. [6], Hills [1], and Furman et al. [7] stated that the relevant target areas for radiotracer applications are defined and that the most appropriate target beneficiaries of radiotracer applications include the mineral processing sectors, petroleum and petrochemical industries, and wastewater treatment plants. These industries are widespread internationally and are of considerable economic and environmental importance. Moreover, according to the IAEA [2, 3], radiotracer techniques have many advantages, such as high detection sensitivity, in-situ detection, availability of a wide range of compatible radiotracers for different phases, rapid response and high reliability, and accuracy of results.

The measurement of homogeneous mixing efficiency is one of the main applications of radiotracers in the industry. Mixing involves the blending of two or more miscible fluids to obtain a predetermined degree of homogeneity. Stirred tanks are widely used in the process industries to perform many different operations, including the blending of miscible liquids into a single liquid phase, the suspension of solids, the promotion of heat and mass transfer, gas-liquid and liquid-liquid mass transfer, crystallisation, and chemical reactions [8, 9]. Several objectives must be fulfilled when a mechanically stirred vessel is used. Some of these objectives include the homogenisation of single or multiple phases at a specific temperature and concentration of components, which can be affected by the physical properties of the fluids that are being mixed.

According to Shekhar and Jayanti [10], the main requirement of the mixing process is to combine two or more fluids that are initially separated. Rahimi and Parvareh [8] observed that, in the liquid phase, the use of impellers and jets are two established methods for fluid homogenisation. Moreover, in the chemical, mineral, and wastewater treatment industries, mechanically stirred tanks are widely used for either simple liquid mixing or for more complex multiphase processes, such as gas-liquid or gas-liquid-solid mixing. Aubin et al. [11] concluded that, to understand the complex phenomena that occur in these mixing tanks, it is necessary to investigate the single- and two-phase flow fields in the vessel and the turbulence characteristics in turbulent applications. Thus, Montante and Magelli [12] suggested an investigation of the flow field that is established in a stirred vessel because it is the most important characteristic of the stirred tank reactor in the processes in which the flow field can affect the homogenisation level. Static mixer, another mixing related service of motionless pipeline devices, is widely used throughout the chemical and hydrocarbon processing industries. This type of mixer is very powerful in the pipeline mixing and is a very dominant option for the laminar flow regime. High reliability over a broad range of flow conditions is achieved when a properly designed static mixer is in operation. Nevertheless, most industrial mixing processes take place in tanks or vessels. Thus, this paper is only concerned with mixing operation in the mixers or vessel.

The residence time distribution (RTD) is one of the important parameters that can provide information on the characteristics of the reactor, such as the flow pattern that occurs [13]. The RTD, which was first developed by Danckwerts [14] has been utilised by many researchers to diagnose possible system malfunctions, such as bypassing, leakage, blockage, channel fouling, and backmixing, and to help estimate the quality of mixing. The RTD, which depends on the flow hydrodynamics and the reactor geometry, influences the chemical reactor performance by affecting a number of reaction properties, such as the conversion and yield. The RTD can be measured by evaluating the concentration of a tracer compound, which is added as a stimulus at the system inlet. A tracer, which has been discussed by Hills [1], the IAEA [2, 3], and Furman et al. [7], has been implemented in experiments that consist of a common impulse-response method in which the tracer is injected at the inlet of the system and the concentration-time curve is recorded at the outlet.

Stegowski and Furman [15] described the fundamentals of the RTD measurement set up, which is illustrated in Figure 2. The RTD curve of a radiotracer experiment is considered measurable after treatment of the raw data. The treatment of the data involves background correction, radioactive decay correction, starting point correction, filtering, and data extrapolation [16]. Moreover, according to Ding et al. [17] the RTD is a fundamental parameter in reactor design because it can provide information on how long the substrate has been in the reactor and it can help characterise the extent of the deviation in the reactor behaviour from ideal condition. Levenspiel [18] mentioned that the quantified RTD can provide a numerical characterisation of the mixing in a reactor, which helps the process engineer better comprehend the mixing performance of the reactor. In addition, Ding et al. [17] agreed that the dimensionless RTD can potentially be used to compare two different equipment designs and operating conditions.

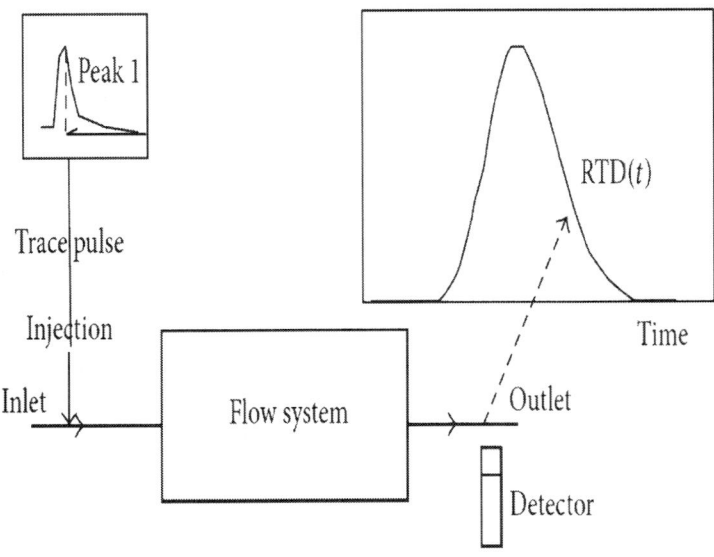

Figure 2: Principle measurement of RTD by Stegowski and Furman (2004) [15].

In addition to the RTD, the mixing time is an important parameter that can be used to determine the homogenisation that occurs in a mixing vessel [16, 19]. There are many definitions of mixing time, which depend on the selected measurement technique. Jafari and Mohammadzadeh [20] defined mixing time as the period of time necessary for a system to achieve the desired level of homogeneity in a given vessel, whereas Bujalski et al. [21] and Patwardhan and Joshi [22] defined mixing time as the time required for the concentration variance to reach a predetermined value. Pramparo et al. [23] and Wang et al. [24] measured the mixing time through the time monitoring evolution of the concentration of a tracer. The tracer can be a chemical species, any substance that can be tracked, or a thermal disturbance, and the measurement techniques that have been used include liquid crystal thermograph, visual observation [21, 25], conductivity, and laser-induced fluorescence [24, 26, 27].

CURRENT PROBLEMS WITH CONVENTIONAL TRACERS

Extensive studies have been conducted to study the efficiency and flow characteristics of mixing vessels using a nonradiotracer approach. The techniques have been successfully implemented and published, but there are some discrepancies when these are compared with radiotracer techniques. Jafari and Mohammadzadeh [20] and Wang et al. [24] measured the mixing time in a vessel by monitoring the injected tracer concentration as a function of time. These researchers found large deviations in the mixing time between the different measurement locations and detection methods used and concluded that the mixing time was dependent on the location and the detection of the tracer. According to them, this phenomenon was due to the removal of small vortexes, which increases the circulation speed of the liquid vortex. Pramparo et al. [23] described the evolution of mixing time as a function of the impeller rotational velocity, as shown in Figure 3, and indicated that the impeller speed is directly proportional to the mixing time.

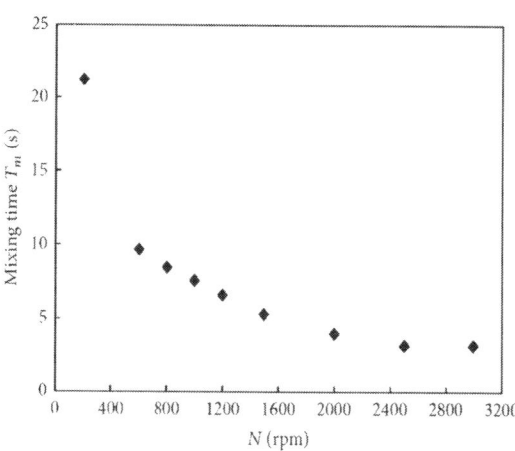

Figure 3: Correlation of the mixing time with the impeller speed by Pramparo et al. 2008 [23].

There is another drawback associated with the use of nonradioactive tracers. Delvigne et al. [27] used a thermal method to calculate the mixing time in which the conductivity probes, or thermocouples, were attached onto the baffles of the vessel. However, due to the size constraint of the probe, which was 0.45 mm in diameter, only a limited area was covered. Therefore, the results failed to represent approximately 90% of the mixing vessel volume [20, 28] and the measured mixing time obtained in this experiment was only 85%. Another technique for the measurement of the mixing time is the use of reagent visualisation. Wabo et al. [25] visualised, in 3D, an acid-alkali reactive tracer in a typical batch stirred vessel reactor, whereas Bujalski et al. [21] calculated the mixing time using the decolourisation of starch. The results of the experiments that were carried out using the thermal method failed to represent approximately 90% of the mixing vessel; only 70% of the fluid volume can be observed through the visualisation by phenolphthalein using a single video camera, as was reported by Wabo et al. [25] and shown in Figure 4. Wabo and his coworkers also used the electrical resistance tomography (ERT) system following the adjacent pair protocol to visualise the fluid mixing during a chemical reaction. They concluded that ERT can suitably image the mixing of reactive tracers and therefore serve as a potential method for quantification of the macrosegregation of reagents in 3D. A total of 16 equally spaced electrodes were used in this study for the current injection; these form a peripheral ring with the 15 voltage pair measurements, as shown in Figure 5. Arratia et al. [13] and Ding et al. [17] concluded that it is inevitable that the introduction of these probes will affect the flow pattern. In addition, the use of these probes involves a complex setup procedure, as described by Wabo et al. [25]. Therefore, the use of a radiotracer is the preferred stimulus response technique because of its noninvasive application and online monitoring systems, which avoid the shutdown of the plant. In addition, Pant and Yelgoankar [19] also declared that radiotracers often have no competing alternative for troubleshooting full-scale industrial reactors. Thus, the integration of radiotracers with computer simulations resolves the previously mentioned problems.

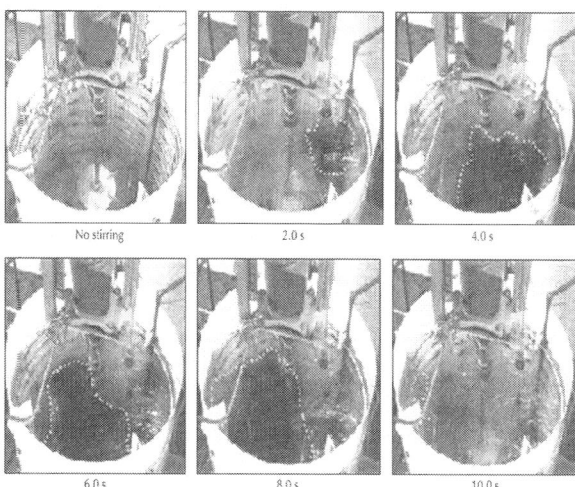

Figure 4: Visualized acid-alkali reactive tracer mixing using the caustic soda-hydrochloric acid-phenolphthalein system conducted by Wabo et al. 2004 [25].

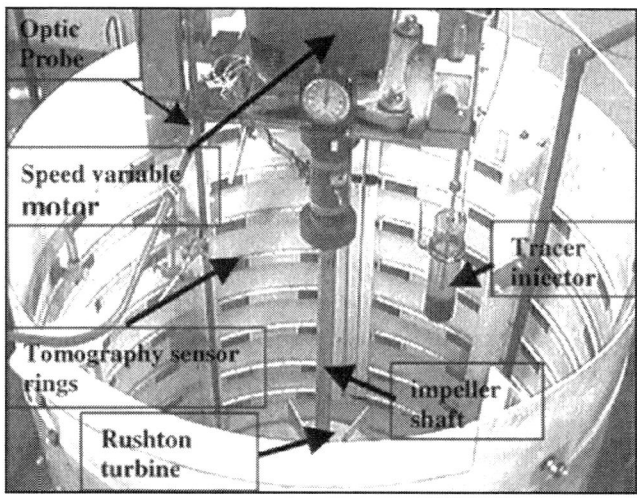

Figure 5: The 8-plane, 16-electrode tomography sensor system fitted for ERT measurement by Wabo et al. (2005) [25].

MATHEMATICAL MODELS: CORRELATION AND DEFINITIONS

Correlation and Definitions: RTD

RTD, or, E(t) is a probability distribution function that describes the amount of time a fluid element spends inside a reactor. It helps in troubleshooting reactors and characterises the macromixing and flow within a reactor [24]. If an impulse of tracer injected at the inlet of a system at time t equals 0 and its concentration is measured as a function of time at the outlet, E(t) represents the probability that a trace element has a residence time between the time interval (t,t+dt) and is defined as follows:

$$E_{out}(t) = \frac{C_{out}(t)}{\int_0^\infty C_{out}(t)\,dt} \text{ such that } \int_0^\infty E_{out}(t)\,dt = 1,$$

(1)

where $C_{out}(t)$ is the detected output signal. The detected signal is normalised by dividing it by the area under the curve, as shown in (1). The mathematical expression for the first moment in discrete form can be written as follows:

$$Mi = \frac{\int_0^\infty t\,C_i(t)\,dt}{\int_0^\infty C_i(t)\,dt}.$$

(2)

Thus, the experimental mean residence time (MRT) of the system is calculated as the difference between the first moments of the outlet and inlet response curves, where i=1 and 2:

$$MRT = M_2 - M_1,$$

(3)

Where M_1 is the moment of the curve monitored at the inlet and M_2 is the moment of the curve monitored at the outlet. Nevertheless, these results should be treated to remove any noises.

Jafari and Mohammadzadeh [20] stated that the RTD can be described by various models, such as a CSTR with exchange volumes, CSTR with a dead zone, and CSTR with a bypass. These models contain many parameters, including the mean residence time, exchange and bypass flow rates, and volume of the dead zone, which can be varied to fit the experimental data. Nevertheless, the models mentioned above do not consider the flow field within the reactor, which results in nonideal behaviour. Levenspiel [18] also noted that there are two commonly cited methods for analysing the RTD curve; these are the dispersion model and the tanks-in-series model. Burrows et al. [29] also addressed more complex methods for the calculation of the volume by short-circuiting the dead zones within the reactor. The dispersion model is based on the ideal plug flow and accounts for the deviations from the ideal flow that is caused by backmixing or random fluctuations. The dispersion number is calculated from the variance of the RTD curve, as shown below:

$$\sigma^2 = 2\frac{D}{uL} - 2\left(\frac{D}{uL}\right)^2 \left(1 - e^{-uL/D}\right),$$

(4)

Where D/uL is the dispersion number (dimensionless),L is the length of the reactor x, is a point along the length of the reactor such that 0<x<L, and t is the mean hydraulic residence time. According to Levenspiel [18] a large dispersion number (greater than 0.2) indicates that the reactor behaviour is similar to a single CSTR, whereas the reactor approximates ideal plug flow if the dispersion number is close to zero. Additionally, Burrows et al. [29] discussed the tanks-in-series model, which can also be used to measure the RTD. This model assumes that the flow through the reactor can be characterised by a series of N equal-sized CSTRs. A tank approaching plug flow would have a large N value, of approximately 30 or higher, whereas the value off N or a completely mixed tank will be one.

Correlations and Definitions: Mixing Time

There are several different correlations that can be used for the prediction of the mixing time in an agitated vessel, which achieve a certain degree of mixture uniformity. Pant et al. (2001) [5] and Bujalski et al. (2002) [21] used a correlation developed by Fasano and Penney (1991) [26] to calculate the mixing time:

$$t_U = \frac{-\ln(1-U)}{1.06N(D/T)^{2.17}(T/H)^{0.5}},$$

(5)

Where U is the degree of uniformity and 0<U<1. Moreover, Pant et al. [5] and Bujalski et al. [21] used the following relationship for t_{95}, which was proposed by Ruszkowski [28]:

$$t_{95} = 5.91T^{-2/3}\left(\frac{\rho V}{P}\right)^{1/3}\left(\frac{T}{D}\right)^{1/3},$$

(6)

where V is the liquid volume in the tank and P is the power input.

Bujalski et al. [21] graphically represented the t_{90} mixing time using the normalised simulated concentration response, as shown in Figure 6. This figure showed that the homogeneity of mixing fluid was obtained at 90% degree of uniformity after tracer injection. The normalized tracer concentration indicates that each collected data were divided by summation of collected data as the denominator. Nevertheless, the values greater than one were due to fluctuations of parasitic signals which were not eliminated or treated earlier as suggested by IAEA 2008 and Kasban et al. prior to RTD determination [3, 16]. Moreover, Pramparo et al. [23] stated that the mean mixing time can be obtained by determining the time between two consecutive peaks from the time-variation curve, as shown in Figure 7. Consider

$$UR_c = \frac{C_\infty - C(i,j,t)}{C_\infty - C_0},$$

$$\theta_{mix} = \frac{1}{m} \sum_{i=1}^{m} \left[\frac{1}{n} \sum_{j=1}^{n} t_{95}(i,j) \right],$$

$$N\theta_{mix} = 5.98 N_{q^{-1/3}} \left(\frac{T}{D} \right)^2.$$

(7)

The above correlations were derived by Zadghaffari et al. [30] to measure the mixing time required for a particular point to reach 95% of the concentration in the tank. However, the direct comparison of the mixing times of two reactors is only possible if the value of Y for the two reactors is the same. Thus, Jafari and Mohammadzadeh [20] suggested that the desired degree of homogeneity should be defined at a convenient value of V (e.g., 0.90 or 0.95). The mixing in the reactor to achieve a specific level of homogeneity can be expressed in terms of the degree of mixing, or Y:

$$Y = \left| \frac{C_{(t)} - C_0}{C_\infty - C_0} \right|,$$

(8)

where C_0 and C_∞ are the initial and final average uniform tracer concentrations, respectively, and $C(t)$ is the tracer uniform concentration in the vessel at time.t.

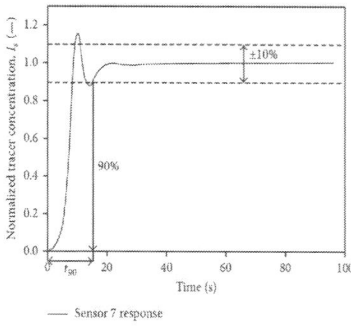

Figure 6: A graphical representation of t_{90} mixing time using the simulated concentration response by Bujalski et al. 2002 [21].

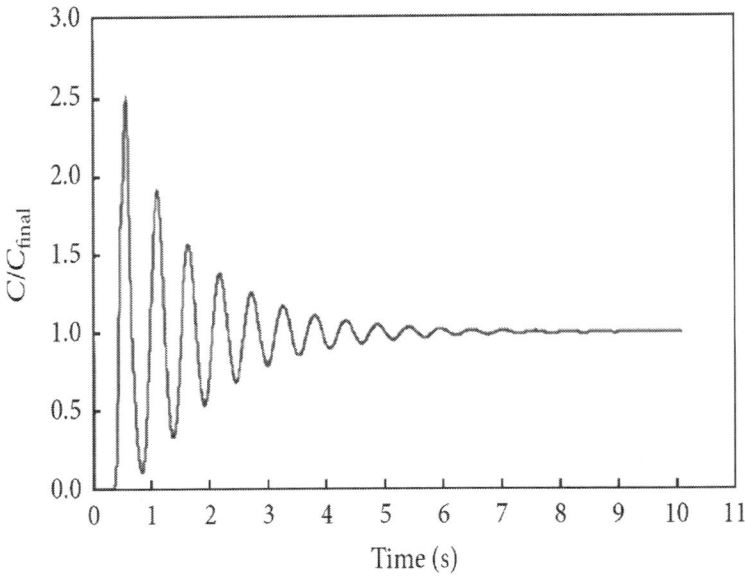

Figure 7: Time evolution of the normalized concentration averaged on the monitor surface by Pramparo et al. 2008 [23].

MATHEMATICAL MODEL IN RADIOTRACER TECHNOLOGY

Extensive radiotracer experiments have been carried out successfully in various industries, which indicate the survival and reliability of radiotracers in hostile environments. To verify its feasibility, the IAEA has developed six mathematical models to analyse the experimentally obtained radiotracer curves. The six proposed models are the axial dispersion plug flow model, the axial dispersed plug flow with exchange model, the perfect mixers in series model, the perfect mixers in series with exchange model, the perfect mixers in parallel model, and the perfect mixers with recycle model [3, 16]. The optimised model curve that best fits the experimental curve is chosen.

A series of radiotracer experiments for the measurement of the RTD, mixing time, and flow rate was conducted by Kasban et al. [16] using $300\,\mu Ci^{99}Mo$ as the radioactive tracer, which was added at a height of 800 mm in the mixing vessel. Four paddle impellers were installed and pretreatment of the raw data was carried out using Matlab. The IAEA software was utilised to characterise the RTD mathematical models that were recommended by the IAEA. The perfect mixer with recycle model best represented the curve of the experimental data and the RTD was found to be 57 min. A flow rate of approximately $8.75\,Ls^{-1}$ was used to accommodate the 175 L of water in the flow rig system. The authors concluded that the speed of the impeller affects the mixing time required for the system to reach homogeneity. However, the authors did not describe the calculation of the mixing time or the malfunction that was identified from the findings. They also did not accurately describe the RTD experimental setup and the mixing time measurement. In addition, the authors did not clarify the impeller size and geometry and the parameters used in the mixing process optimisation.

Radiotracers have also successfully assisted industries in the research of multiphase flow. Sugiharto et al. [31] determined the RTD and the system flow rate in a 24 in. multiphase flow hydrocarbon transmission pipeline containing approximately 95% water, 3% crude oil, 2% gas, and negligible solid material. Nevertheless, the types of radiotracer sources used in this experiment were different; I-13 and Na-24 were used independently for the measurement of the RTD in hydrocarbons and water, respectively. In this instance, the tanks-in-series model best described the RTD of the system. The authors also discovered that the water moved faster than the hydrocarbon even though the density of the water is higher. This might be because water is more dominant in the transmission line and because the movement of the crude oil is slowed by friction with gas at the top layer and friction at the water-crude oil interface. Moreover, Behin and Aghajari [32] studied the RTD measurement in a pilot-scale oil-water separator operated by Drood oil of the Iranian Offshore Oil Company (IOOC) located in Kharg Island using 5mCi I-131 as sodium iodide and 4mCi of I-131 as iodobenzene for the aqueous and organic phase, respectively. The separation of the crude oil and water mixtures is an important process in the oil and chemical industries. The researchers reported that the experimental results were in good agreement using the model of perfect mixing tanks-in-series (with a dead zone) to describe the liquid behaviour. The models obtained in these two case studies were both tanks-in-series models. It can therefore be concluded that tanks-in-series models suited the multiphase flow profile.

The next case studies, which were conducted by Pant et al., indicate the superiority of radiotracer technology in hostile industrial environments. In addition, each case study developed a mathematical model that fit the experimental results satisfactorily although the chosen models varied between applications. Pant et al. [33] conducted an RTD study in a pilot and an industrialised fluid catalytic cracking unit (FCCU) using an intrinsic tracer. The identified tracers were lanthanum-140 and sodium-24 and were obtained from the catalyst sample. The tracers were characterised using neutron activation analysis to investigate the degree of axial mixing and radial distribution in the riser section of the FCCUs and to determine the residence time distribution of the catalyst. The axial dispersion model (ADM) was used to represent the actual RTD obtained from the radiotracer experiments. According to Pant et al., the ADM is the best model to describe material flow in a tabular reactor, such as FCCU, because, according to Kasban et al. [16] this model can describe one-dimensional convection and dispersion in a pipe. Moreover, Pant et al. [6] measured the RTD of coal particles in a pilot-scale fluidised bed gasifier (FBG). These researchers successfully used 100 g of each lanthanum-140- and gold-198-labelled coal particles or 100 g of lanthanum-140- and gold-198-labelled coal particles (50 g of each) at a very high temperature (1000°C) as the radiotracer source. They represented the behaviour of the coal particles that flow from the bottom of the gasifier with tanks-in-series model. The results revealed that there was a good degree of mixing and that only a small fraction of the feed material bypassed and short-circuited from the bottom of the gasifier. This was also the first report of the use of radiotracers in an FBG in India and proved the superiority of the radiotracer techniques in the energy industry.

Pant et al. [5] also determined the RTD in a sludge hygienisation research irradiator (SHRI) using 7–10 mCi NH_4^{82} Br as the tracer to troubleshoot the malfunction in the reactor. Their results disclosed the presence of a dead zone in approximately one-fourth of the irradiator with low flow rates and a semistagnant volume and slow flow exchange with higher flow rates. In this study, the tracer first appearance time (TFAT) was implemented to estimate the minimum dose received by the sludge from one circulation through the irradiator. The authors discussed the methods used to determine the flow rates, circulation time, and homogenisation time (mixing time) from the derived data, as shown in Figure 8. The three models used to measure these three reactor characteristics were the models of ideal mixers in series with plug flow in a recirculation loop, ideal mixers in series with plug flow inside the irradiator and in a recirculation loop, and ideal mixers in series with stagnant volume and plug flow in a recirculation loop, respectively. All of these model simulations provided important information about the hydrodynamic behaviour of the closed recirculation, batch-type sludge irradiator. In addition, the case studies that were conducted by Pant et al. showed the ability of radiotracer technology not only to survive in hostile industrial environments but also to troubleshoot and pinpoint the malfunctions or anomalies in the plants of different industries. The good agreement that was obtained between the developed models and the experimental results indicate the accuracy of the theories.

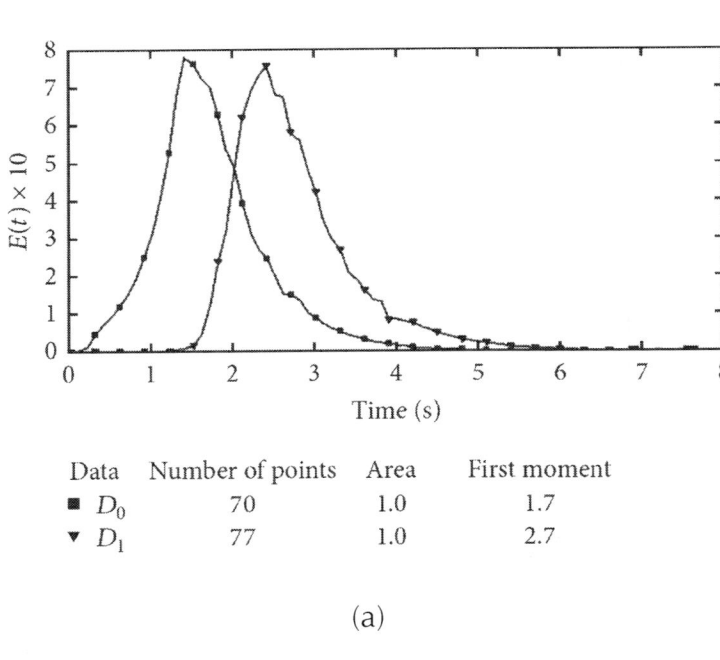

Data	Number of points	Area	First moment
■ D_0	70	1.0	1.7
▼ D_1	77	1.0	2.7

(a)

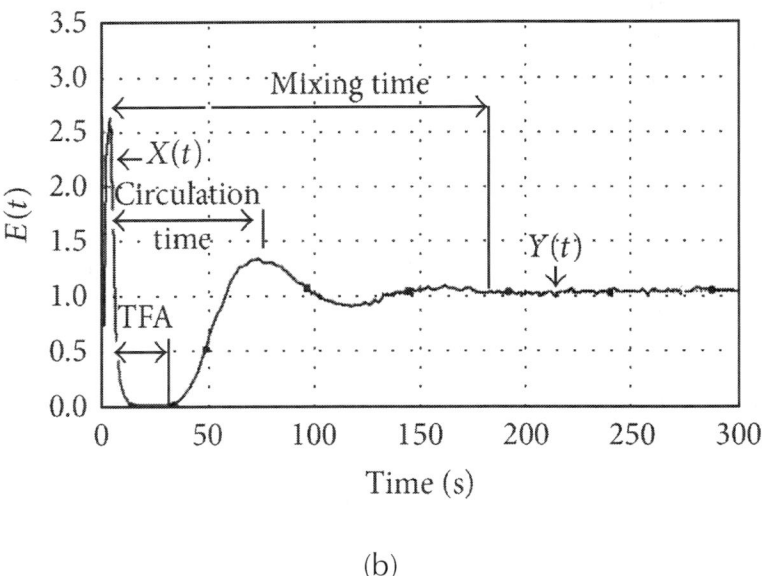

(b)

Figure 8: (a) Flow rate measurement; (b) time of first appearance, circulation, and homogenisation or mixing time by Pant et al. 2001 [5].

Moreover, Pant and Yelgoankar [19] investigated the RTD in servotherm special oil as the heat transfer medium (HTM) in two identical aniline production reactors, one of which was the reference reactor. These researchers used [82]Br as para-dibromo benzene as the radioactive source. The output from the RTD indicated the presence of undesired parallel flow streams in the shell-side of the abnormal reactor due to a 60% fouling of the reactor. The data were treated prior to any RTD analysis to ensure that only the radioactive material was analysed. The authors implemented tanks-in-series model to simulate the results and detected the presence of several anomalies in the reactor, which were mainly fouling/scaling or dead volume. Due to the large amount of fouling in the reactor and the possibility of the occurrence of parallel streams in the shell-side of the reactor, they implemented a model of tanks-in-series with two parallel streams to represent the RTD curve that was obtained from the radiotracer experiments. The results showed the ability of the radiotracer technology to highlight the percentage of abnormalities that were present in the processing plant accurately, which was impossible to achieve with other conventional methods.

The next case studies demonstrate that radiotracer technology can be used in any form, that is, solid, gas, or liquid. The form of the radiotracer source is based upon the medium of the plant under investigation to obtain superior physicochemical compatibility. Santos and Dantas [34] determined the RTD by calculating the transit of the methylbromide gaseous tracer and the nondispersive Co-60 tracer in an FCC cold mode. They found good agreement between the Experimental Cold Model (ECM) and the RTD results obtained from the radiotracer experiments. Moreover, Klusener et al. [35] investigated the RTD in a horizontal cross-flow bubble column reactor of a commercial ethylbenzene oxidation plant using Ar-41 gas as the tracer. These researchers implemented the axial dispersion and the tanks-in-series models to describe the mean residence time, the number of tanks N, and the Peclet number. The results showed that the ADM described the gas radiotracer experiments better than the tanks-in-series model. This result makes sense because the ADM describes one-dimensional convection and dispersion in a pipe more accurately than the tanks-in-series model. Thus, the RTD results enabled the plant to implement extensive changes to optimise the plant yields and improve the selectivity at a lower reaction temperature.

Meanwhile, Furman et al. [4] conducted a radiotracer experiment on a ball mill in which the grinding of copper ore occurs. In this study, they used the radioactive isotopes ^{64}Cu and NH_4Br (^{82}Br) as the tracers for the copper sulphurs and water, respectively. They suggested a series of perfect mixers with dead volume to represent the radiotracer experiments. However, in this case study, it was necessary to use a different model to accurately fit the experimental data from the radiotracer experiment. Therefore, the optimised model consists of a serially connected plug flow reactor and perfect mixers with dead volume.

Stegowski and Furman [15] highlighted different RTD models that are commonly used to represent flow patterns of industrial systems in the Laplace s-domain, as shown in Table 1. They measured the RTD in a copper ore thickening, filtration, and drying system using 1 GBq of Cu-64 as the radioactive tracer. The RTD study shows the presence of a dead zone and bypasses, which affect the efficiency of the dewatering system, and provides a detailed analysis of the solid state behaviour inside the dewatering subsystems. The parameters involved for each model are highlighted in the table. In this case study, the main parameters for the plug flow and perfect mixing series model are the residence time, the mean residence time, and the number of perfect mixers.

Table 1: Elementary models used in experimental RTD data analysis [4]

3	Model's symbol	Parameters	Laplace's transformations
Plug flow		T	exp(-Ts)

Perfect mixing cells in series		τ, j	$\left(1+\dfrac{\tau}{j}s\right)^{-j}$
Perfect mixing cells in series with backmixing		$\tau, j\ \alpha,$	$\dfrac{1}{\alpha}\left(1+2\alpha+\dfrac{\tau s}{j}-b\right)$ $b=\left[1+\left(\dfrac{2\tau s}{j}\right)(1+2\alpha)+\left(\dfrac{\tau s}{j}\right)^2\right]^{1/2}$
Perfect mixers in series with exchange zones		$\tau, j, t_m\ K$	$\left[\dfrac{1+t_m s}{1+(t+t_m+tK)s+tt_m s^2}\right]^{j}$ $t=\dfrac{\tau}{j}(1+K)$

The next case study shows the capability to integrate the mathematical RTD model with another model and indicates that the models recommended by the IAEA do not have to be used independently to describe a process but can be integrated easily with other models. In addition, the model can also be used to describe the present anomalies in the reactor or system under study. Klusener et al. [35] studied the influence of the inlet positions on the flow behaviour inside a photoreactor using $1mCi^{113m}$ In. Initially, the flow behaviour was modelled using the small plug flow and perfect mixing cell model, which represents the jet effect. Nevertheless, to represent the whole system, the optimised model consisted of two parameters: the mean residence time in the plug flow reactor and the mean residence time in the perfect mixing cell. The optimised models described the experimental results satisfactorily. To simulate the lateral inlet configurations, the author used the plug flow with axial dispersion model, which uses the mean residence time and the Peclet number as the parameters. The results showed that there was a large stagnant volume in the central inlet, which was reflected in the estimation of the first moments of the RTD curves. According to Nigam et al. [36] new data on backmixing and mass transfer parameters have been obtained from the residence time distribution (RTD) of a liquid in an air-water concurrent down flow trickle-bed reactor that contains six different types of alumina-support porous catalysts, that is, spheres, tablets, holed tablets, and extrudates. They used 10–20 MBq of sodium pertechnatate salt samples containing radioactive ^{99m}Tc to conduct the RTD studies. They implemented a comprehensive axial dispersion exchange and intraparticle diffusion RTD model to interpret the effect of the gas and liquid input flow rates on the liquid axial dispersion coefficients, the liquid-liquid mass transfer coefficients between the stagnant and dynamic liquid zones, and the liquid-solid mass transfer coefficients between the static liquid and the subtended wetted fractional pellet areas.

The typical use of radiotracer approaches is to compare the experimental curves with the RTD mathematical models that are proposed by the IAEA. Nevertheless, there are some papers that do not mention the mathematical model that best describes the radiotracer experiments. Instead, they simply directly highlight the malfunction of the investigated system or vessel. Oriol et al. [37] investigated the feasibility of using the gas radiotracer Krypton-81 m to diagnose the flow maldistribution in a multiphase heat exchanger in which the horizontal pipe was fed by an air-water mixture. Both smooth and perturbed curves were observed throughout the experiments. The curve shape indicates whether the two-phase flow regime is continuous, which can be stratified, annular or wavy, or dispersed, which is represented by plug, slug, or bubbles. The studies indicated that the shape of the signal can be used to determine the gas-liquid regime. Very high frequencies are associated with a bubble flow regime, whereas smaller frequencies are associated with slug or plug flow regimes. Moreover, Abellon et al. [38] studied the residence time or solid circulation rate of a single particle in an interconnected fluidised bed facility (IFB) with glass beads using a glass particle labelled with radioactive^{24}Na or ^{92}Ir. The ^{92}Ir was produced by melted glass and the addition of 0.05 wt% iridium. In this study, the radiotracer particle size was varied to determine the correlation between the radiotracer particle size and the MRT while all other parameters were held constant. The researchers found that the MRT of the radiotracer is independent of its size as long as it is within the size range of the batch. They also concluded that all of the particles in the IFB acted as one "homogenous fluid" and fluidised from one cell to another. The previously described models for the different case studies have not provided insights into the processes that occur in the vessels or systems. The industries are mostly interested in understanding the reasons that underline the flow pattern that arises in the plant instead of simply knowing the mathematical model that would represent it or a graphical profile of the flow. Thus, the use of computational fluid dynamics (CFD) would complement the results obtained from the radiotracer experiments.

COMPUTATIONAL SIMULATIONS USING CFD

Computational fluid dynamics (CFD) is a computer modelling tool that enables the visualisation of complex processes. According to the IAEA [2] and Ranade [39], CFD is a superior and predictive analysis that provides clear spatial pictures of the process under study, which includes information such as flow patterns and velocity maps. Moreover, a number of industrial reactor engineering applications utilise commercial CFD tools to ensure enhanced maintainability. The ability to visualise and monitor the flow pattern in a chemical process plant results in a better understanding of the actual chemical process. Therefore, this increased ability to monitor flow patterns will promote the use of radiotracer technology in many industries, especially a number of Malaysian industries.

RTD in CFD

Numerous experimental and CFD studies on the RTD in tanks have been carried out and published over the years. Zadghaffari et al. [30]

concluded that the experimental validation of the numerical RTD is more convenient than the experimental validation of the complete flow that is obtained numerically. Wang et al. [24], Bai et al. [40], and Furman and Stegowski [41] agreed that the present numerical method used to calculate the mixing time and the residence time distribution in a stirred tank has proven to be reliable and widely applicable. Moreover, the liquid flow and tracer transport can be described by a general partial differential equation in the cylindrical coordinate system in terms of the velocity components,u,v,w the turbulent kinetic energy k, the viscous dissipation ε, and the tracer concentration.c Consider

$$\frac{\partial(\rho\varphi)}{\partial t} + \frac{1}{r}\frac{\partial}{\partial r}(\rho u r \varphi) + \frac{1}{r}\frac{\partial y}{\partial \theta}(\rho v \varphi) + \frac{\partial(\rho w \varphi)}{\partial z}$$

$$= \frac{1}{r}\frac{\partial}{\partial r}\left(r\Gamma_{\phi,\mathrm{eff}}\frac{\partial\phi}{\partial r}\right) + \frac{1}{r}\frac{\partial}{\partial\theta}\left(\frac{\Gamma_{\phi,\mathrm{eff}}}{r}\frac{\partial\phi}{\partial\theta}\right)$$

$$+ \frac{\partial}{\partial z}\left(\Gamma_{\phi,\mathrm{eff}}\frac{\partial\phi}{\partial z}\right) + S, \tag{9}$$

Where φ is the dependent variable and S is the source term per volume. The eddy diffusivity was modelled using a standard $k - \varepsilon$ model and the impeller rotation was modelled using the improved inner-outer iterative procedure [17] with multiple reference frames [24, 29] and a sliding mesh approach [42]. Bai et al. [40] used the realizable $k - \varepsilon$ model and multiple reference frames (MRFs) to simulate the stirred tank at steady-state. Furman and Stegowski [41] obtained the reactor RTD from stochastic particle tracking or discrete phase models and used a large number of tracers ($\sim 10^5$) to reduce the statistical uncertainty. They concluded that both the standard and the RNG k- models predict the experimental RTD successfully, but a significant deviation was obtained between the simulated and experimentally measured mean residence time and tank space time, which indicates that the stochastic tracking model failed to accurately reproduce the flow behaviour [40].

There are several factors that influence the mean residence time. In a CFD model, the rotational speed does not directly affect the mean residence time. Nevertheless, experimental evidence shows that the mean residence time decreases with increasing rotational speed. The centrifugal force that is generated by a high rotational speed results in the appearance of dead zones near the wall surfaces of the vessel and most of the fluid flow moves directly along the stirring shaft to the outlet. However, as the inlet flow rate increases, a change in the rotational speed does not affect the mean residence time. Cao et al. [43] explained that, to quantitatively understand the RTD, the experimental data have to be fitted to an adequate model, as described in Section 3, to describe the nonideal reactor flow pattern, which will provide a better understanding of the quality of mixing, especially for reactions that are not first order. Moreover, the placement of a stirrer at the bottom of an annular reactor with an optimal speed of 150–250 rpm could narrow the RTD curve and improve the mass transfer for surface reactions. The reactor increasingly behaved as a single stirred tank reactor, which is reflected by the broadening of the RTD curve at higher stirring rates. Furman and Stegowski [41] described an example of curve fitting the RTD data with other models, as shown in Figure 9.

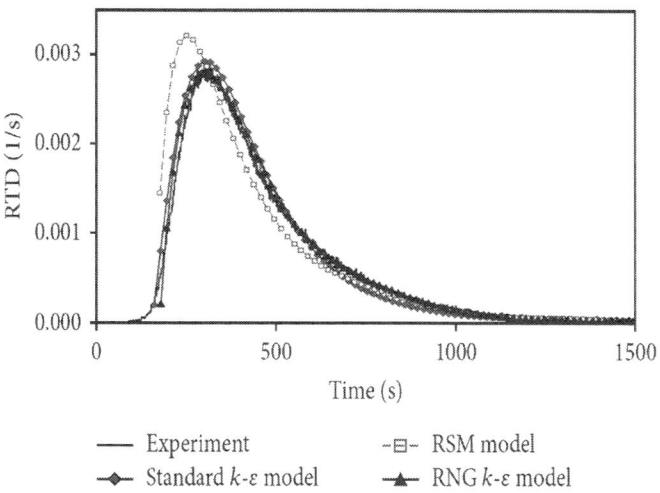

Figure 9: Experimental and simulated RTD functions by Furman and Stegowski 2011 [41].

The detailed description of a reactor malfunction using RTD data is best described by Cao et al. [43] as shown in Figure 10. The $E(\theta)$ curves show the presence of stagnant zones. The appearance of the maximum $E(\theta)$ at $\theta < 1$ indicates the presence of short-circuits, whereas the appearance of the maximum $E(\theta)$ at $\theta > 1$ indicates the presence of backmixing. However, if the maximum of the $E(\theta)$ curve appears near $\theta = 1$ at an impeller speed of $50\,r\,min^{-1}$, the flow pattern approximates ideal plug flow, as illustrated in Figure 10(g). However, if the maximum of $E(\theta)$ is offset to $\theta < 1$, the flow is influenced by the presence of short-circuits at an impeller speed of $400\,r\,min^{-1}$. Sahle-Demessie et al. [44] and Gavrilescu and Tudose [45] concluded that the width of the RTD curve is an appropriate measurement to determine the approximation of the flow pattern to plug flow.

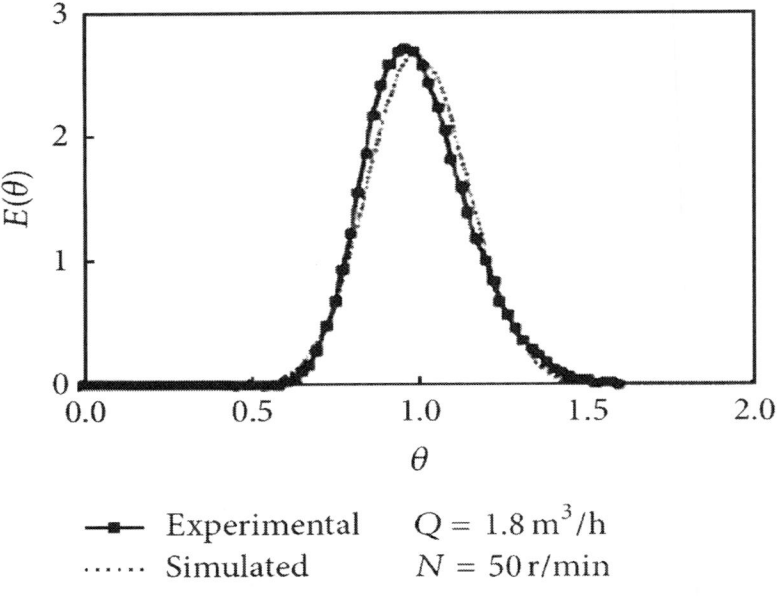

(a)

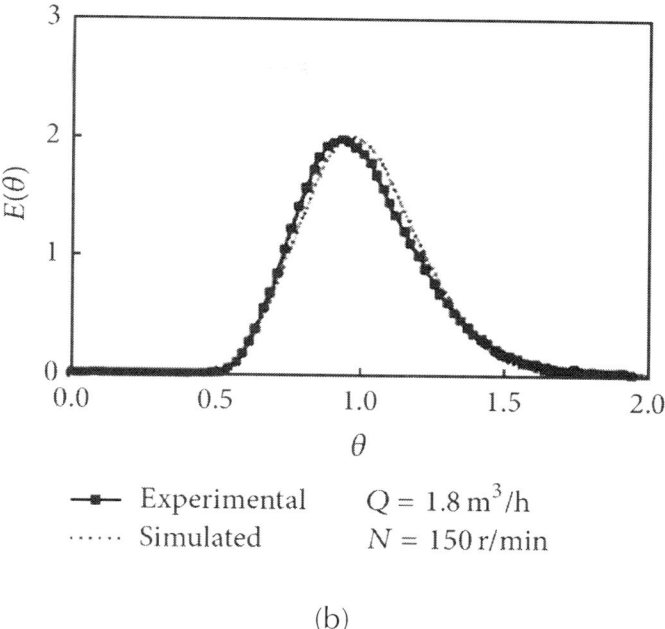

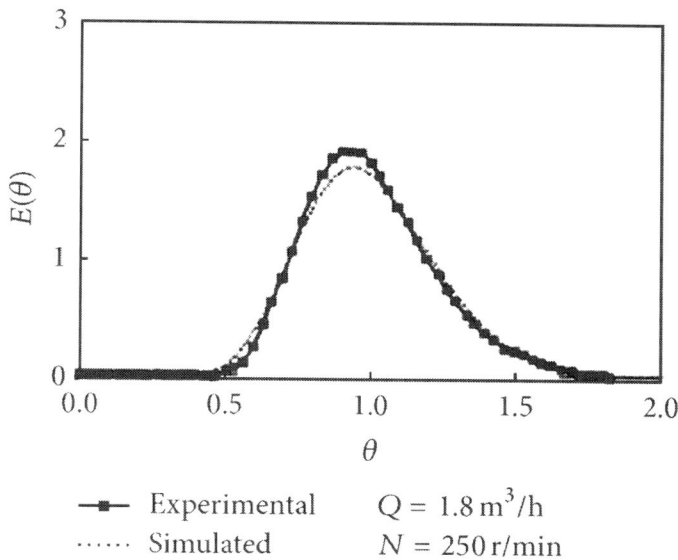

(c)

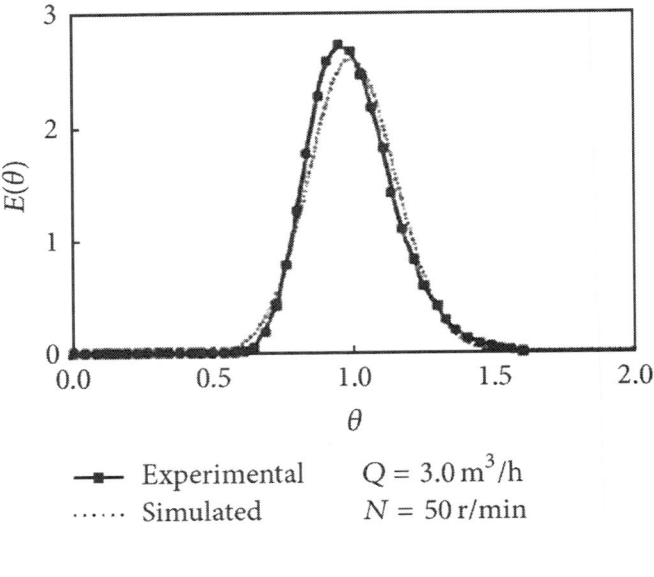

(d)

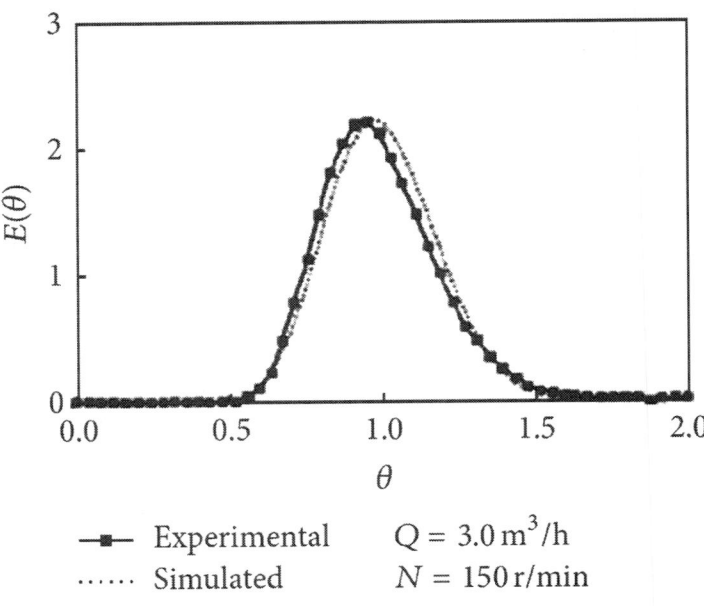

(e)

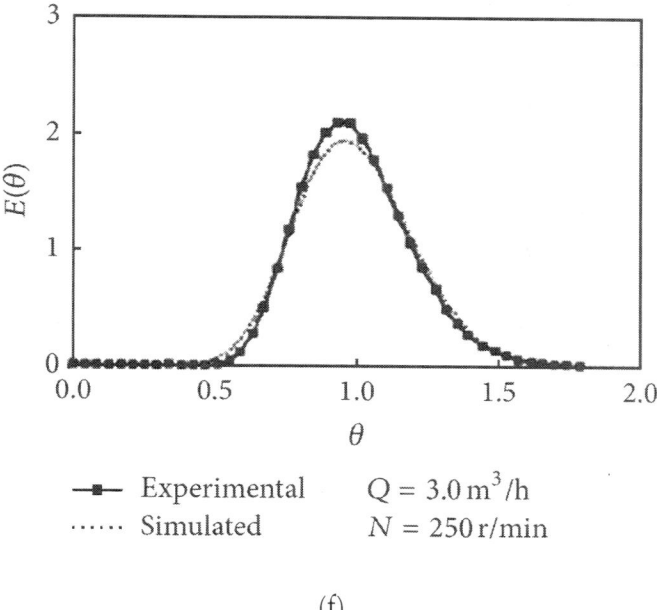

(f)

(g)

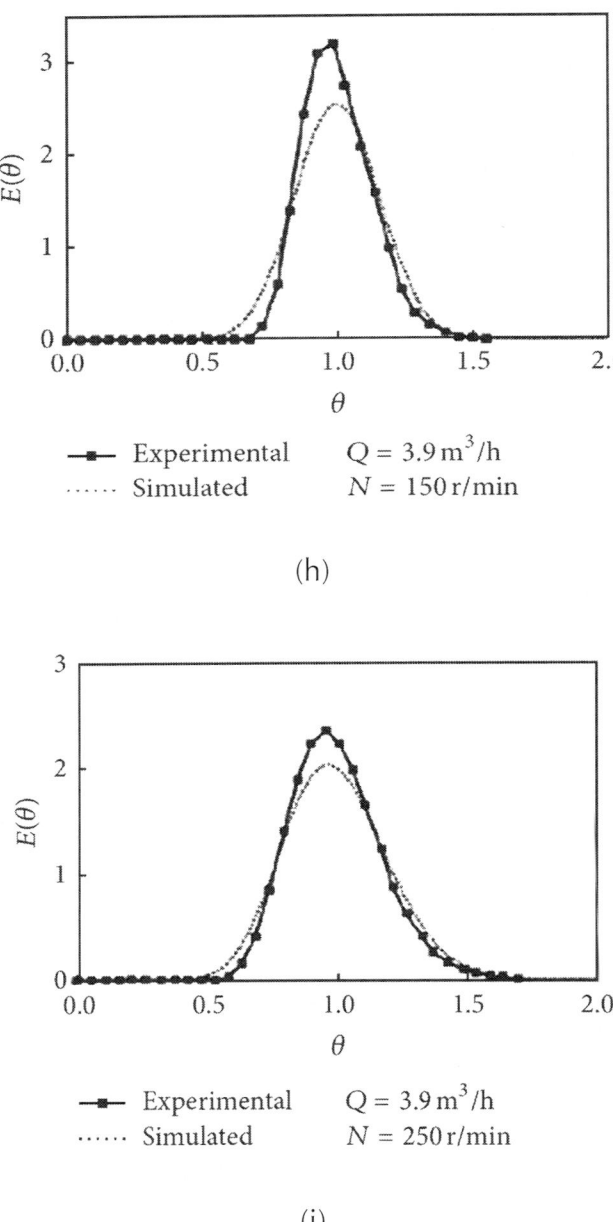

(h)

(i)

Figure 10: Example of reactor malfunction from RTD curves by Cao et al. 2009 [43].

Another factor that affects the RTD is stirring. Cao et al. [43] explained that the RTD curve is narrow with stirring, which causes the flow to approach plug flow conditions. However, although slow stirring can narrow the RTD curve and improve the performance of the reactor, increasing the impeller speed widens the RTD curve. Figure 11 shows the numerical RTD curves at different impeller speeds. From the figure, it can be concluded that the RTD curve tends to flatten as the speed of the impeller increases, which indicates that a high impeller speed can also worsen reactor performance. Thus, researchers should conduct preliminary experiments to identify a set of reliable impeller speeds prior to process optimisation. Jaffari and Mohammazadeh [20] studied the effect of the liquid flow rate on the RTD in a gas-induced contactor with a constant impeller speed and liquid volume of 1100 rpm and 10 litres, respectively. Because of the constant impeller speed, it was expected that the values of the maximum concentration [C(t)] and the corresponding times would be equal. The RTD curve became wider with decreasing liquid flow rates, which increased the mean residence time, as shown in Figure 12. Hence, by increasing the impeller speed, the time corresponding to the maximum point of the RTD curves decreases. It can therefore be concluded that the impeller speed and the liquid flow rate can have a significant effect on the RTD curve.

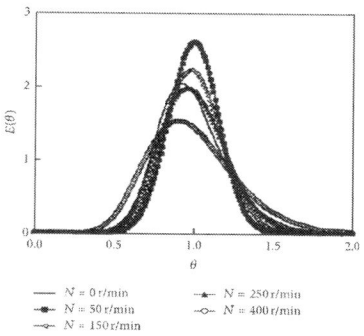

Figure 11: Effects of impeller speed with RTD curves by Cao et al. 2009 [43].

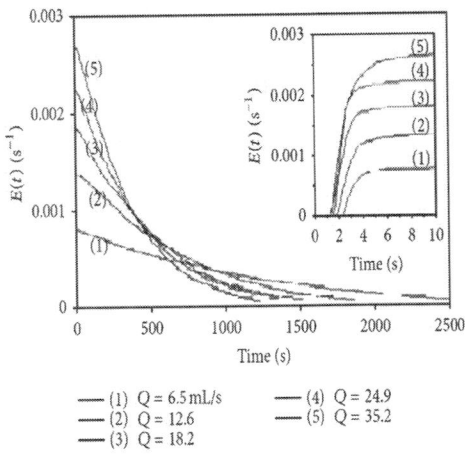

Figure 12: Effects of flow rate with RTD [20].

Mixing Time in CFD

Numerical simulation is another technique that is used to calculate mixing time. CFD is currently widely used to verify experimental results. Osman and Varley [46] measured mixing time in an unbaffled vessel with a Rushton turbine using the moving reference frame (MRF) approach and Jaworski et al. [42] studied homogenisation in a baffled vessel stirred by a dual Rushton impeller using a similar approach. They both predicted a mixing time that was 2-3 times higher than the measured value and attributed this deviation to a wrongly predicted tangential velocity field and an under-prediction of the mass exchange between the recirculation zones that are generated by the turbines. Thus, Bujalski et al. [21] used the transient scalar transport equation in a stationary reference frame for a baffled reactor; however, the mixing time was still over-predicted by a factor of approximately two, which is similar to the results obtained by Jaworski et al. [42]. To directly compare their results with the study performed by Osman and Varley [46] the authors should also attempt to apply the scalar transport equation to an unbaffled vessel. They should also use the low Reynolds inmodel, which was successfully used by Shekhar and Jayanti [10] to simulate the flow field and mixing characteristics in an unbaffled vessel stirred with a paddle impeller. The studies conducted by Zadghaffari et al. [47] who used the MRF solution as the starting point and switched to a sliding mesh for the unsteady conditions and Bujalski et al. [21] who implemented a sliding mesh approach concluded that the mixing time depends on the feed point because the inner rotating mesh volume, which was used to model the rotating impeller, was the main factor that affected the distribution of the tracer.

Yianneskis [48] reported that the position of the probe and the tracer injection point do not have a significant effect on the final homogenisation time. Rahimi and Parvareh [8] discovered that the prediction of the mixing using the RNG version of the k- model yielded better results with realisable and standard conditions, whereas Zadghaffari et al. [47] discovered that the use of a sliding mesh technique with the LES turbulent model was faster than the RNG model, but over-predicted the liquid homogenisation time. Nevertheless, Javed et al. [49] used a fully predictive sliding mesh technique and a secondary liquid as a tracer and found that the discrepancy between the predicted and the actual transport of the tracer within a vertical plane was caused by the location where the tracer injection was simulated. Zadghaffari et al. [30] and Zadghaffari et al. [47] agreed that, at given and constant inlet and outlet flow rates, the mixing efficiency decreased at higher agitation speeds because the stronger radial outflow and increasing Reynolds number push the species rapidly into the lower and upper recirculation loops. Hence, the impeller rotational speed increases the liquid pumping capacity of the impeller and thus reduces the mixing time.

HYBRID OF RADIOTRACER AND CFD

The use of CFD modelling in process industries has attracted a lot of attention since 1990. Modelling and numeric simulation have been used to validate a large amount of experimental data, especially in process industries. Rahimi and Parvareh [8] stated that this technique was developed due to the availability of advanced measurement techniques that could be used to validate the theoretical results. Recently, tracer engineers have started to integrate radiotracer experiments with computer simulation to improve the industrial process visualisation and optimisation. The objective of the integration of radiotracer experiments and visualisation modelling is to provide insights into the industrial process and to ensure that the movement of the radiotracer is within safe boundaries. Moreover, the technology provided by the CFD simulations allows the monitoring and tracking of a radiotracer, which will increase the confidence of using radiotracers in plants. Thus, this combination of technologies can provide a significant contribution to the industries because it can be used to localise any anomalies, process malfunctions, or mechanical damages in a plant without requiring its shutdown. Hence, expensive downtime is avoided and the convenience of direct measurements results in lower process investigation costs. Although there is limited literature on the hybrid of radiotracer technology and CFD, a large amount of research is currently being carried out in most IAEA countries on the development of this technology.

Stegowski and Furman [15] used a solution of $Na^{99m}TcO_4$ as the radioactive source for the investigation of the RTD in a laboratory jet mixer. They compared the experimental RTD derived from the radiotracer experiments with the simulations from three different CFD models: standard k- , RNG k- , and RSM. The authors found that the RNG k- , model best represents the radiotracer experiment curve. However, the authors did not describe the type of RTD mathematical model that they utilised. The authors also failed to highlight the anomalies in the laboratory jet mixer. Another hybrid investigation was conducted by Din et al. [50] who integrated CFD modelling with radiotracer experiments in a two phase counter-current pulsed sieve plate extraction. They used 0.5 mCi Tc-99m and the axial dispersion model to determine the RTD and to study the hydrodynamics of the dispersed phase. They used the

RTD data to estimate the proportions of water and kerosene as a continuous phase. They discovered that the axial dispersion model represents approximately 72.17% of the experimental value but it is still acceptable for the study of the hydrodynamics. The CFD models used were multiphase, standard k-ε, porous media, and pulse generation models. In their paper, the authors described the type of model used, as recommended by the IAEA. However, the results obtained from the CFD simulation were not acceptable because they deviated approximately 30% from the experimental measurements. It is possible that another type of model, such as the RNG or realisable standard k–ε models, might yield better results.

The combination of radiotracer technology and computer modelling has increased the use of both of these technologies, which indicates its superiority among the conventional techniques that are used worldwide.

CONCLUSIONS

In this review, the development of radiotracer technology has been described. The introduction described radiotracer technology in general, which was followed by a summary of the current problems that arise from the use of conventional tracers. A description of the mathematical models that are used in radiotracer applications was then presented, followed by a summary of computer simulations to analyse the RTD and mixing time and an introduction of a hybrid of radiotracer technology and computational fluid dynamics (CFD). The literature described in this review indicates the versatility and feasibility of radiotracer technology in numerous industrial and laboratory applications. Future research on radiotracer technology will fully utilise and validate its use with numerical solutions in the study of mixing vessels because radiotracer techniques are widely used for the identification of industrial malfunctions and measurement of process parameters, such as the mean residence time (MRT), residence time distribution (RTD), and flow rate. Radiotracer technology has many advantages over conventional tracers; these include its high detection sensitivity, physicochemical compatibility, in-situ detection, and limited memory effect. It is possible that radiotracer technology can be used in combination with an experimental design that implements the Taguchi Orthogonal Array Method, which is a robust design method that was developed to reduce cost and improve the quality of chemical process [51]. The Taguchi method has been applied initially to determine the minimum number of experiments that are required for process optimisation, which is beneficial because of the inclusion of radioactive experiments. This method involves the use of orthogonal array techniques to investigate the simultaneous variation of several parameters and the interactions between parameters.

ACKNOWLEDGMENTS

The authors gratefully acknowledge the financial support by the Malaysian Nuclear Agency, Ministry of Science, Technology and Innovation (MOSTI) under Grant NM-R&D-11-25 and Science Fund 03-03-01-SF0075.

REFERENCES

1.   A. E. Hills, Practical Guidebook for Radioisotope-Based Technology in Industry, IAEA/RCA RAS/8/078, 2nd edition, 2001.

2.   Radiotracer Technology as Applied to Industry, IAEA-TECDOC-1262, IAEA, Vienna, Austria, 2001.

3.   IAEA, "Residence time distribution method for industrial and environmental applications," Training Course Series 31, Vienna, Austria, 2008.

4.   L. Furman, L. Petryka, Z. St gowski, and A. Wierzbicki, "Data acquisition and processing in radiotracer experiments," Nuclear Instruments and Methods in Physics Research B, vol. 211, no. 3, pp. 436–442, 2003.

5.   H. J. Pant, J. Thýn, R. Zitný, and B. C. Bhatt, "Radioisotope tracer study in a sludge hygienization research irradiator (SHRI)," Applied Radiation and Isotopes, vol. 54, no. 1, pp. 1–10, 2001.

6.   H. J. Pant, V. K. Sharma, M. V. Kamudu et al., "Investigation of flow behaviour of coal particles in a pilot-scale fluidized bed gasifier (FBG) using radiotracer technique," Applied Radiation and Isotopes, vol. 67, no. 9, pp. 1609–1615, 2009.

7.   L. Furman, J. P. Leclerc, and Z. Stegowski, "Tracer investigation of a packed column under variable flow," Chemical Engineering Science, vol. 60, no. 11, pp. 3043–3048, 2005.

8.   M. Rahimi and A. Parvareh, "Experimental and CFD investigation on mixing by a jet in a semi-industrial stirred tank," Chemical Engineering Journal, vol. 115, no. 1-2, pp. 85–92, 2005.

9.   H. J. Pant, J. Thýn, R. Zitný, and B. C. Bhatt, "Radioisotope tracer study in a sludge hygienization research irradiator (SHRI)," Applied Radiation and Isotopes, vol. 54, no. 1, pp. 1–10, 2001.

10.   S. M. Shekhar and S. Jayanti, "CFD study of power and mixing time for paddle mixing in unbaffled vessels," Chemical Engineering Research and Design, vol. 80, no. 5, pp. 482–498, 2002.

11.   J. Aubin, D. F. Fletcher, and C. Xuereb, "Modeling turbulent flow in stirred tanks with CFD: the influence of the modeling approach, turbulence model and numerical scheme," Experimental Thermal and Fluid Science, vol. 28, no. 5, pp. 431–445, 2004.

12.   G. Montante and F. Magelli, "Liquid homogenization characteristics in vessels stirred with multiple Rushton turbines mounted at different spacings: CFD study and comparison with experimental data,"Chemical Engineering Research and Design, vol. 82, no. 9, pp. 1179–1187, 2004.

13.   P. E. Arratia, J. P. Lacombe, T. Shinbrot, and F. J. Muzzio, "Segregated regions in continuous laminar stirred tank

reactors," Chemical Engineering Science, vol. 59, no. 7, pp. 1481–1490, 2004.

14. P. V. Danckwerts, "Continuous flow systems. Distribution of residence times," Chemical Engineering Science, vol. 2, no. 1, pp. 1–13, 1953.

15. Z. Stegowski and L. Furman, "Radioisotope tracer investigation and modeling of copper concentrate dewatering process," International Journal of Mineral Processing, vol. 73, no. 1, pp. 37–43, 2004.

16. H. Kasban, O. Zahran, H. Arafa, M. El-Kordy, S. M. S. Elaraby, and F. E. Abd El-Samie, "Laboratory experiments and modeling for industrial radiotracer applications," Applied Radiation and Isotopes, vol. 68, no. 6, pp. 1049–1056, 2010.

17. J. Ding, X. Wang, X.-F. Zhou, N.-Q. Ren, and W.-Q. Guo, "CFD optimization of continuous stirred-tank (CSTR) reactor for biohydrogen production," Bioresource Technology, vol. 101, no. 18, pp. 7005–7013, 2010.

18. O. Levenspiel, Chemical Reaction Engineering, John Wiley & Sons, New York, NY, USA, 3rd edition, 1999.

19. H. J. Pant and V. N. Yelgoankar, "Radiotracer investigations in aniline production reactors," Applied Radiation and Isotopes, vol. 57, no. 3, pp. 319–325, 2002. ·

20. M. Jafari and J. S. S. Mohammadzadeh, "Mixing time, homogenization energy and residence time distribution in a gas-induced contactor," Chemical Engineering Research and Design, vol. 83, no. 5 A, pp. 452–459, 2005.

21. W. Bujalski, Z. Jaworski, and A. W. Nienow, "CFD study of homogenization with dual Rushton turbines—comparison with experimental results. Part II: the multiple reference frame," Chemical Engineering Research and Design, vol. 80, no. 1, pp. 97–104, 2002.

22. A. W. Patwardhan and J. B. Joshi, "Relation between flow pattern and blending in stirred tanks,"Industrial and Engineering Chemistry Research, vol. 38, no. 8, pp. 3131–3143, 1999.

23. L. Pramparo, J. Pruvost, F. Stüber et al., "Mixing and hydrodynamics investigation using CFD in a square-sectioned torus reactor in batch and continuous regimes," Chemical Engineering Journal, vol. 137, no. 2, pp. 386–395, 2008.

24. Z. Wang, Z.-S. Mao, and X.-Q. Shen, "Numerical simulation of macroscopic mixing in a Rushton impeller stirred tank," The Chinese Journal of Process Engineering, vol. 6, no. 6, pp. 857–863, 2006.

25. E. Wabo, M. Kagoshima, and R. Mann, "Batch stirred vessel mixing evaluated by visualized reactive tracers and electrical tomography," Chemical Engineering Research and Design, vol. 82, no. 9, pp. 1229–1236, 2004.

26. J. B. Fasano and W. R. Penney, "Avoid blending mix-ups," Chemical Engineering Progress, vol. 87, no. 10, pp. 56–63, 1991.

27. F. Delvigne, J. Destain, and P. Thonart, "Structured mixing model for stirred bioreactors: an extension to the stochastic approach," Chemical Engineering Journal, vol. 113, no. 1, pp. 1–12, 2005.

28. S. Ruszkowski, "A rational method for measuring blending performance, and comparison of different impeller types," in Proceedings of the 8th European Conference on Mixing, pp. 283–292, Cambridge, UK, 1994.

29. L. J. Burrows, A. J. Stokes, J. R. West, C. F. Forster, and A. D. Martin, "Evaluation of different analytical methods for tracer studies in aeration lanes of activated sludge plants," Water Research, vol. 33, no. 2, pp. 367–374, 1999.

30. R. Zadghaffari, J. S. Moghaddas, and J. Revstedt, "A mixing study in a double-Rushton stirred tank,"Computers and Chemical Engineering, vol. 33, no. 7, pp. 1240–1246, 2009.

31. S. Sugiharto, Z. Suʼud, R. Kurniadi, W. Wibisono, and Z. Abidin, "Radiotracer method for residence time distribution study in multiphase flow system," Applied Radiation and Isotopes, vol. 67, no. 7-8, pp. 1445–1448, 2009.

32.   J. Behin and M. Aghajari, "Influence of water level on oil-water separation by residence time distribution curves investigations," Separation and Purification Technology, vol. 64, no. 1, pp. 48–55, 2008.

33.   H. J. Pant, V. K. Sharma, A. G. C. Nair et al., "Application of ^{140}La and ^{24}Na as intrinsic radiotracers for investigating catalyst dynamics in FCCUs," Applied Radiation and Isotopes, vol. 67, no. 9, pp. 1591–1599, 2009.

34.   V. A. Santos and C. C. Dantas, "Transit time and RTD measurements by radioactive tracer to assess the riser flow pattern," Powder Technology, vol. 140, no. 1-2, pp. 116–121, 2004.

35.   P. A. A. Klusener, G. Jonkers, F. During et al., "Horizontal cross-flow bubble column reactors: CFD and validation by plant scale tracer experiments," Chemical Engineering Science, vol. 62, no. 18–20, pp. 5495–5502, 2007.

36.   K. D. P. Nigam, I. Iliuta, and F. Larachi, "Liquid back-mixing and mass transfer effects in trickle-bed reactors filled with porous catalyst particles," Chemical Engineering and Processing, vol. 41, no. 4, pp. 365–371, 2001.

37.   J. Oriol, J. P. Leclerc, P. Berne et al., "Characterization of two-phase flow regimes in horizontal tubes using 81 mKr tracer experiments," Applied Radiation and Isotopes, vol. 66, no. 10, pp. 1363–1370, 2008.

38.   R. D. Abellon, Z. I. Kolar, W. Den Hollander, J. J. M. De Goeij, J. C. Schouten, and C. M. Van Den Bleek, "A single radiotracer particle method for the determination of solids circulation rate in interconnected fluidized beds," Powder Technology, vol. 92, no. 1, pp. 53–60, 1997.

39.   V. V. Ranade, Computational Flow Modeling for Chemical Reactor Engineering, Academic Press, New York, NY, USA, 2002.

40.   H. Bai, A. Stephenson, J. Jimenez, D. Jewell, and P. Gillis, "Modeling flow and residence time distribution in an industrial-scale reactor with a plunging jet inlet and optional

agitation," Chemical Engineering Research and Design, vol. 86, no. 12, pp. 1462–1476, 2008.

41. L. Furman and Z. Stegowski, "CFD models of jet mixing and their validation by tracer experiments,"Chemical Engineering and Processing, vol. 50, no. 3, pp. 300–304, 2011.

42. Z. Jaworski, W. Bujalski, N. Otomo, and A. W. Nienow, "CFD study of homogenization with dual rushton turbines-comparison with experimental results. Part I: initial studies," Chemical Engineering Research and Design, vol. 78, no. 3, pp. 327–333, 2000.

43. X.-C. Cao, T.-A. Zhang, and Q.-Y. Zhao, "Computational simulation of fluid dynamics in a tubular stirred reactor," Transactions of Nonferrous Metals Society of China, vol. 19, no. 2, pp. 489–495, 2009.

44. E. Sahle-Demessie, S. Bekele, and U. R. Pillai, "Residence time distribution of fluids in stirred annular photoreactor," Catalysis Today, vol. 88, no. 1-2, pp. 61–72, 2003.

45. M. Gavrilescu and R. Z. Tudose, "Residence time distribution of the liquid phase in a concentric-tube airlift reactor," Chemical Engineering and Processing, vol. 38, no. 3, pp. 225–238, 1999.

46. J. J. Osman and J. Varley, "Use of computational fluid dynamics (CFD) to estimate mixing times in a stirred tank," in Proceedings of the 6th Fluid Mixing Symposium, Institution of Chemical Engineers Symposium Series, pp. 15–22, August 1999.

47. R. Zadghaffari, J. S. Moghaddas, and J. Revstedt, "Large-eddy simulation of turbulent flow in a stirred tank driven by a Rushton turbine," Computers and Fluids, vol. 39, no. 7, pp. 1183–1190, 2010.

48. M. Yianneskis, "The effect of flow rates and tracer insertion time on mixing times inkjet-agitated vessels," in Proceeding of the 7th European Conference on Mixing, pp. 121–128, Brugge, Belgium, 1991.

49. K. H. Javed, T. Mahmud, and J. M. Zhu, "Numerical simulation of turbulent batch mixing in a vessel agitated by a Rushton turbine," Chemical Engineering and Processing, vol. 45, no. 2, pp. 99–112, 2006.

50. G. U. Din, I. R. Chughtai, M. H. Inayat, I. H. Khan, and N. K. Qazi, "Modeling of a two-phase countercurrent pulsed sieve plate extraction column—a hybrid CFD and radiotracer RTD analysis approach," Separation and Purification Technology, vol. 73, no. 2, pp. 302–309, 2010.

51. A. R. Khoei, I. Masters, and D. T. Gethin, "Design optimisation of aluminium recycling processes using Taguchi technique," Journal of Materials Processing Technology, vol. 127, no. 1, pp. 96–106, 2002.

Chapter 4

Perpendicular Magnetic Anisotropy in FePt Patterned Media Employing a CrV Seed Layer

Hyunsu Kim[1], Jin-Seo Noh[1], Jong Wook Roh[1], Dong Won Chun[2], Sungman Kim[2], Sang Hyun Jung[3], Ho Kwan Kang[3], Won Yong Jeong[2], and Wooyoung Lee[1]

[1]Department of Materials Science and Engineering, Yonsei University, Seoul 120-749, Korea

[2]Korea Institute of Science and Technology (KIST), Seongbuk-gu, Seoul 136-761, Korea

[3]Nano Process Division, Korea Advanced Nano Fab. Center, Gyeonggi 443-270, Korea

ABSTRACT

A thin FePt film was deposited onto a CrV seed layer at 400°C and showed a high coercivity (~3,400 Oe) and high magnetization (900–1,000 emu/cm^3) characteristic of $L1_0$ phase. However, the magnetic properties of patterned media fabricated from the film stack were degraded due to the Ar-ion bombardment. We employed a deposition-last process, in which FePt film deposited at room temperature underwent lift-off and post-annealing processes, to avoid the exposure of FePt to Ar plasma. A patterned medium with 100-nm nano-columns showed an out-of-plane coercivity fivefold larger than its in-plane counterpart and a remanent magnetization comparable to saturation magnetization in the out-of-plane direction, indicating a high perpendicular anisotropy. These results demonstrate the high perpendicular anisotropy in FePt patterned media using a Cr-based compound seed layer for the first time and suggest that ultra-high-density magnetic recording media can be achieved using this optimized top-down approach.

INTRODUCTION

Conventional planar magnetic recording methods have been facing difficulties in reducing the thickness of a magnetic film and the average grain size in it, which is required for the high bit density [1, 2]. Furthermore, these methods showed a bit density limit of about 100 Gbit/in^2 due to the magnetic moment instability termed 'superparamagnetism' in very small grains and the magnetic exchange interaction between adjacent grains [1-4]. To overcome this limit, a perpendicular magnetic recording was introduced, where magnetic moments are aligned perpendicular to the film plane [5, 6]. However, a bit loss still occurs by exchange interaction between neighboring grains. Patterned magnetic media have emerged as a means to prevent this intergranular exchange interaction, thus to achieve the ultra-high density of magnetic recording. To realize the very fine patterned media, a proper material stack and well-optimized fabrication process should be chosen to retain the magnetization in the perpendicular direction with a high perpendicular anisotropy (Ku).

FePt is a magnetic material that has been intensively investigated due to its high coercivity (Hc = 1–10 kOe) [7-11] and high magnetocrystalline anisotropy (Kc = 7.0 × 10^7 erg/cm^3) [8,10, 12]. This material undergoes a transition from chemically disordered face-centered cubic phase (FCC, $A1$ phase) to ordered face-centered tetragonal phase (FCT, $L1_0$ phase) at a specific temperature, and the transition temperature and perpendicular anisotropy are known to depend on the buffer layer and process employed. A variety of buffer layers have been introduced on Si or glass substrates to grow high quality FCT structures at low temperatures, including Pt, Au, Ag, Ti, and MgO [13-16]. Although $L1_0$ FePt films on these buffer layers demonstrated an increase in coercivity with respect to the buffer-free films, the ratio of out-of-plane to in-plane coercivities has generally been smaller than 3. Other than these rather conventional buffer layers, Cr-based compounds such as CrW [17] and CrRu[18] have also been examined as underlayers since (200) planes of a body-centered cubic (BCC) Cr were likely to stimulate (001) texture formation of the FCT FePt and to facilitate the FCC-to-FCT transition in FePt layer by forcing the tensile stress to a_0 side of the original FCC FePt [17,18], achieving the ratio of out-of-plane to in-plane coercivities larger than 5 at a relatively low temperature (400°C) [18]. To our knowledge, however, no works have successfully demonstrated the high perpendicular anisotropy in FePt fine-patterned media employing a Cr compound, presumably due to the difficulty in optimal process design.

In this work, we fabricated magnetic recording media by a combination of E-beam lithography and either dry etching (deposition-first process) or lift-off (deposition-last process), where magnetic nano-columns were regularly arranged with a fixed spacing. The magnetic properties and crystal structures were investigated at important steps of the fabrication of the patterned media. The high perpendicular anisotropy is demonstrated in the fine-patterned media, suggesting the feasibility of achieving the ultra-high-density recording media through a well-designed fabrication process.

EXPERIMENTAL

A 70-nm-thick CrV seed layer was sputter-deposited at 400°C on a glass substrate. Then, a FePt layer 7 nm in thickness was deposited on top of the CrV at 400°C by ultra-high vacuum (UHV, 3×10^{-8} Torr) sputtering [19]. Patterned media were fabricated from this film stack, following the conventional top-down process (deposition-first process) shown in Figure 1a. In this process, a type of negative E-beam resist (ER), hydrogen silsesquioxane (HSQ), was used for E-beam lithography. The coated ER was baked at 110°C for 60 s before E-beam irradiation. Going through E-beam exposure and development in tetramethylammonium hydroxide (TMAH), regular ER columns were patterned: typical diameter and pitch of the ER patterns were 100 and 200 nm, respectively. Using the ER patterns as etch masks, the inductively coupled plasma (ICP) Ar etching was performed for 1 min under 15 sccm of Ar flow to transfer the ER patterns onto the film stack. The etching was stopped right below FePt/CrV interface. The ER was finally removed, leaving behind FePt patterns, as shown in the last panel of Figure 1a.

As an alternative process, a lift-off process (deposition-last process) was employed to fabricate the patterned media, as shown in Figure 1b. For this process, a type of positive ER was coated on CrV layer and patterned undergoing E-beam exposure and development steps, leaving behind a regular array of holes of a fixed size (typically, 100 nm). Then, a 7-nm-thick FePt layer was deposited by sputtering at room temperature, followed by a lift-off. The final FePt patterns (the last panel of Figure1b) were subsequently annealed at 400°C for 1 h to induce a phase transformation from $A1$ to $L1_0$ phase.

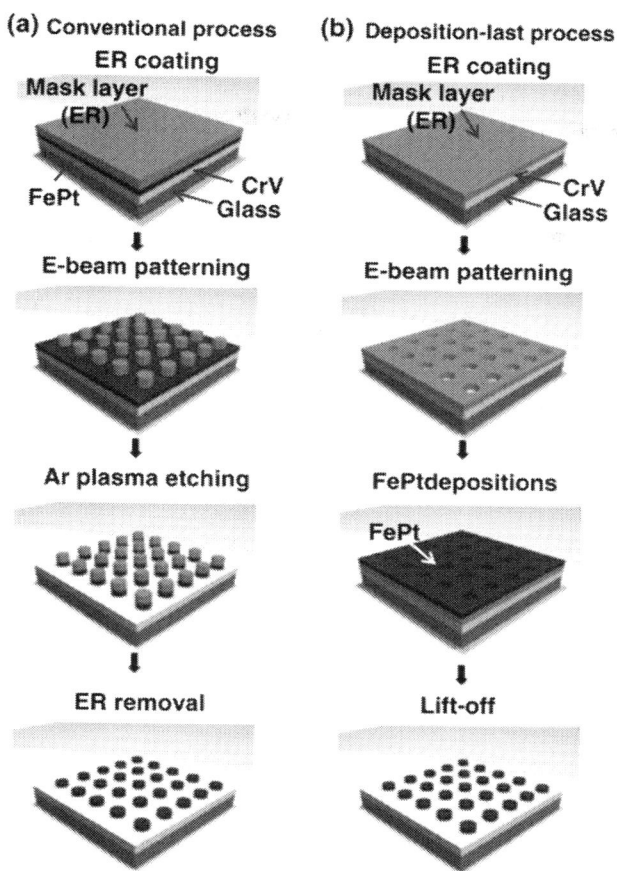

Figure 1: Schematic pictures showing fabrication procedures of FePt patterned media: (a) conventional top-down process and (b) deposition-last process.

To analyze the crystal structures of as-grown films and patterned media, conventional θ–2θ X-ray diffraction (XRD) was performed using Cu $K\alpha$ radiation. Magnetic properties were investigated at room temperature, using a superconducting quantum interference device (SQUID) with a sensitivity of 1×10^{-6} emu. Microstructures of the film stacks and top-views of the fabricated patterned media were observed using transmission electron microscopy (TEM) and scanning electron microscopy (SEM), respectively.

RESULTS AND DISCUSSION

Figure 2a shows a TEM image of an as-grown FePt/CrV film stack. The CrV seed layer exhibits a well-developed columnar grain structure. From our previous study, the well-defined columnar grains of the CrV layer was found to induce perpendicularly oriented grains in a thin FePt overlayer, which resulted in $L1_0$ FePt film at a moderate temperature [20]. To confirm this, we performed a XRD measurement on the as-grown FePt/CrV film stack. As seen in Figure 2b, characteristic FCT (001) and (002) peaks are observed without any FCC peaks, indicating that the FePt film is really in the $L1_0$ phase.

The noisy baseline and rather broad FePt peaks are probably due to the very small thickness (7 nm) of the FePt film. Using this $L1_0$ FePt film on a CrV seed layer, FePt patterned media were fabricated. Figure 2cshows the FePt patterns of different sizes (100 and 50 nm in diameter) fabricated by the combined use of E-beam lithography and Ar plasma etching. The FePt nano-columns having a circular cross section are regularly arrayed on the CrV/glass substrates. The spacing between neighboring nano-columns is the same as its diameter, making the pitch a twofold of the diameter (200 and 100 nm for the respective pattern). From the figure, it is apparent that FePt patterns down to 50 nm in size (100 nm in pitch) can be fabricated by our top-down approach. As a matter of fact, we confirmed that the pattern size could be reduced to 25 nm with 50 nm pitch. Below this size limit (25 nm), the nano-columns started to be deformed, leading to a partly connected array.

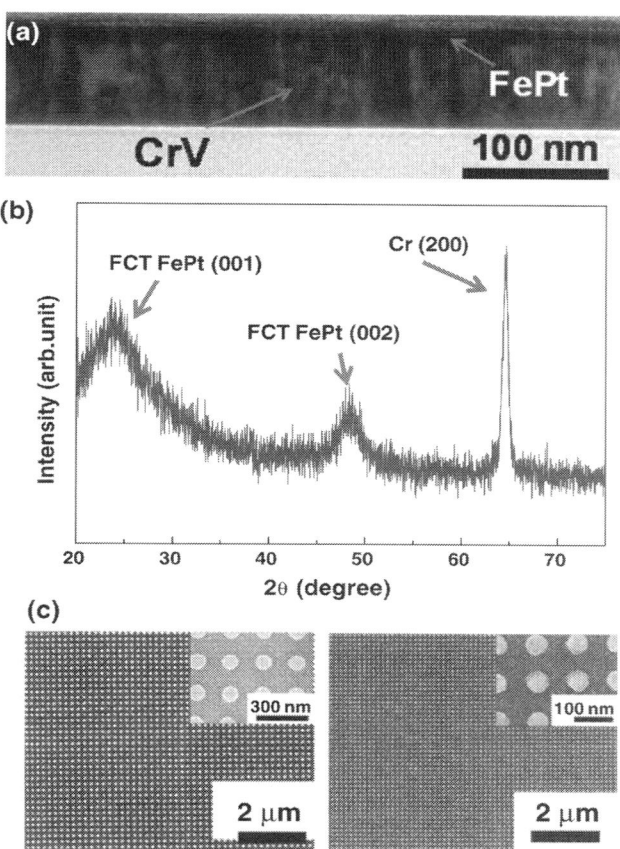

Figure 2: (a) Transmission electron microscopy (TEM) image of thin FePt layer on columnar CrV seed layer. (b) XRD pattern of the as-grown film stack. (c) Scanning electron microscopy (SEM) images of FePt patterns of 100 nm diameter (left) and 50 nm diameter (right), respectively. *Insets*show magnified views of the respective patterns for clarity.

We carried out magnetic field sweepings on the patterned media to investigate the magnetization (M) versus magnetic field (H) behaviors of the media, using a SQUID. Figure 3a shows the M versus Hloops measured at room temperature for the as-grown FePt film (out-of-plane) and a patterned medium with 100-nm-sized columns (both in-plane and out-of-plane). The saturation magnetization ($M_{s,film}$) and coercivity ($H_{c,film}$) of the as-grown film are

900–1,000 emu/cm^3 and ~3,400 Oe, respectively, which are close to those previously reported for FePt $L1_0$ phase [8]. These values and the high ratio of remanent magnetization ($M_{r,film}$) to saturation magnetization, $M_{r,film}/M_{s,film} \approx 1$, may be another indicators that the FePt film was ordered into $L1_0$ phase during deposition at 400°C. It is believed that the formation of complete $L1_0$ phase at a temperature lower than widely adopted post-annealing temperatures (500–800°C) [15,21-23] is attributed to both the high surface diffusivity of adatoms at the elevated deposition temperature and good morphology transfer from the CrV seed layer to a growing FePt film.

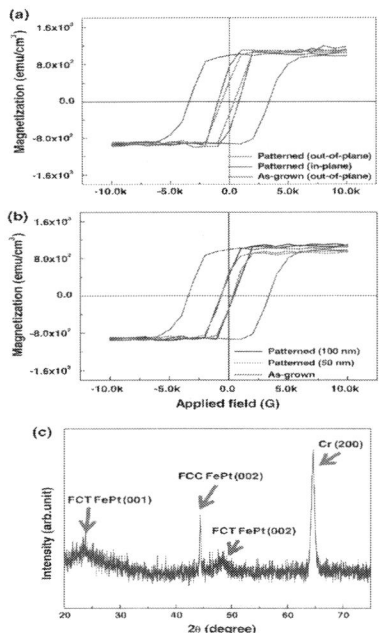

Figure 3: (a) M vs. H curves at room temperature for the as-grown film (out-of-plane) and a patterned medium with 100-nm-sized columns (out-of-plane and in-plane). (b) Out-of-plane M vs. H curves for the as-grown film and patterned media with 100 and 50 nm columns. (c) XRD pattern of a patterned medium with 100-nm-sized columns.

However, the coercivities ($H_{c,pattern} = 450$–900 Oe) of the patterned medium appear to be 4 to sevenfold smaller than $H_{c,film}$ both in film plane and normal to plane, although its saturation magnetizations ($M_{s,pattern}$) are similar to $M_{s,film}$. In addition, the ratio ($M_{r,pattern}/M_{s,pattern} = 0.4–0.7$) of $M_{r,pattern}$ to $M_{s,pattern}$ for the medium is smaller than that of the as-grown film. Recollecting that the coercivity and M_r/M_s ratio are more structure-sensitive than the saturation magnetization, these results suggest that the chemically ordered FCT structure was destroyed and replaced by the chemically disordered FCC structure at least partially during ICP Ar etching. To verify this presumption, we carried out the XRD analysis on the patterned medium. Indeed, it is shown from Figure 3c that the FCT (002) peak is weak and instead, a FCC (002) peak is clearly developed around $2\theta = 44.5°$, justifying the propriety of the above presumption. We think that the large decrease in coercivity for the patterned medium originated from the relaxation of magnetocrystalline anisotropy (Kc) due to the chemical disordering in the FePt patterns [7,24,25]. This is because shape anisotropy ($Kd \alpha \alpha M2s$, where α is the demagnetization factor) strengthens the perpendicular alignment of magnetic moments, and magnetoelastic anisotropy (K where is the magnetostriction constant and is the stress in film) remains almost unchanged via patterning [26]. The Ar-ion penetration into the FePt film and a large momentum delivered from impinging Ar ions may be primary sources for the collapse of the FCT structure. The drastic decrease in coercivity was also observed in other patterned media with different pattern size, as shown in Figure 3b. It is seen from this figure that both coercivities and M_r/M_s ratios for patterned media are significantly reduced from the values of the as-grown film irrespective of pattern size, reflecting the FCT structure was collapsed for samples undergoing Ar plasma etching as confirmed by the XRD result in Figure 3c.

To avoid this direct exposure of FePt film to Ar plasma, we modified the fabrication procedure of patterned media as illustrated in Figure 1b. Based on this deposition-last process, the FePt film remains intact because no ion impingement is involved in whole fabrication steps. Figure 4a shows the FePt patterns produced by a combination of E-beam lithography and FePt lift-off. The FePt patterns of 100 nm size (200 nm pitch) are circular in shape and uniformly spaced from their neighbors, making pattern quality comparable to that of the top-down patterns mentioned above (see Figure 2c for comparison). A XRD measurement on the deposition-last patterned medium confirms that this modified process allows for realization of the $L1_0$ phase in fine-patterned FePt, as seen from Figure 4b. Magnetic hysteresis loops for this deposition-last patterned medium are shown in Figure 4cfor both applied field directions of out-of-plane and in-plane. Now, a perpendicular anisotropy is clearly observed, making the direction perpendicular to film plane a magnetic easy axis. The coercivities in out-of-plane and in-plane directions are approximately 3,000 and 600 Oe, respectively, resulting in $H_{c,out}/H_{c,in} \approx 5$ for this patterned medium. The strong perpendicular magnetic anisotropy is also supported by the perfect squareness ($M_{r,out}/M_{s,out} \approx 1$) of M-H curve in the out-of-plane direction, while this ratio falls to a half ($M_{r,in}/M_{s,in} = 0.52$) in film plane.

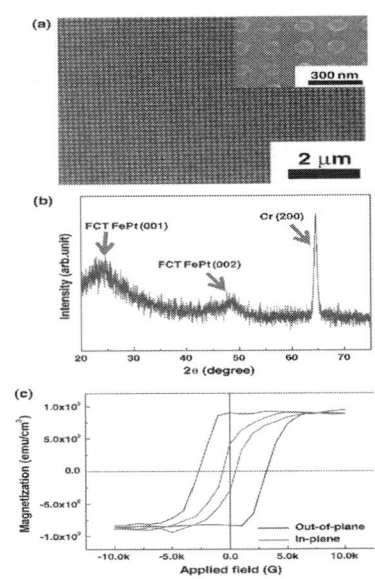

Figure 4: (a) SEM image and (b) XRD pattern of FePt patterns of 100 nm diameter fabricated by the deposition-last process. The *inset* in ashows a magnified view of the pattern for clarity. (c) Comparison of M vs. Hcurves for the patterned medium in out-of-plane and in-plane directions.

Comparing the out-of-plane coercivity of this patterned medium with that of the as-grown film prepared by the deposition-first process, there exists a small difference of about 400 Oe. We believe that this magnitude of difference is reasonable since the surface migration of adatoms during film growth at elevated temperature (400°C) is easier compared to solid-state diffusion of constituents during post-annealing at the same temperature. Qiu et al. also fabricated FePt patterned media with underlayers such as Ag and MgO, employing a similar deposition-last process [15]. In their media, however, the FCC-to-FCT phase transition was retarded to higher temperatures and no perpendicular anisotropy was observed. Assuming that the magnetocrystalline anisotropy is a primary source of our perpendicular anisotropy as explained above, the perpendicular anisotropy is proportional to the coercivity and saturation magnetization in the out-of-plane direction, $KuH_{c,out}M_{s,out}$. In our 100-nm-sized FePt patterns fabricated by the deposition-last process, the values are measured to be 3,000 Oe and 870 emu/cm^3, respectively. These values are comparable to those of the previously reported FePt thin films on other Cr-based compounds such as CrW [17] and CrRu [18], demonstrating the high efficiency of the CrV seed layer in fabricating patterned media with a high perpendicular anisotropy. Besides this, our results disclose important implications: (1) a root cause of the magnetic property degradation of FePt patterned media fabricated by a conventional deposition-first process is chemical disordering incurred by ion plasma etching. (2) The deposition-last process is desirable for implementing ultra-high-density patterned media, and the post-annealing temperature can be maintained low by the support of an appropriate seed layer.

CONCLUSIONS

We fabricated FePt-based perpendicular patterned media using a selective combination of E-beam lithography and either Ar plasma etching (deposition-first process) or FePt lift-off (deposition-last process). A FePt film on a CrV seed layer grown at 400°C showed a high perpendicular anisotropy indicating $L1_0$ phase of FCT structure formed during deposition, whereas the anisotropy was collapsed in patterned media fabricated from the film stack. We employed the deposition-last process to avoid chemical and structural disordering by impinging Ar ions. For a patterned medium with 100 nm patterns made by this process, the out-of-plane coercivity was measured to be fivefold larger than its in-plane value and the out-of-plane M-H curve exhibited a perfect squareness, indicating a high perpendicular anisotropy. To our knowledge, this is the first demonstration of a high perpendicular anisotropy in patterned media using a Cr-based compound seed layer. Furthermore, the deposition-last process may be a promising way to achieve ultra-high-density patterned media due to its maintainability of perpendicular anisotropy and controllability of pattern size and shape.

ACKNOWLEDGMENTS

This research was supported by a grant from the Fundamental R&D Program for Core Technology of Materials funded by the Ministry of Knowledge Economy, Republic of Korea, and the Priority Research Centers Program (2009-0093823) funded by the National Research Foundation of Korea (NRF).

REFERENCES

1.   Moser A, Takano K, Margulies DT, Albrecht M, Sonobe Y, Ikeda Y, Sun S, Fullerton EE: *J Phys D Appl Phys.* 2002, 35:R157.

2.   Albrecht M, Rettner CT, Best ME, Terris BD: *Appl Phys Lett.* 2003, 83:4363.

3.   Bertram HN, Zhou H, Gustafson R: *IEEE Trans Magn.* 1998, 34:1845.

4.   Acharya BR, Inomata A, Abarra EN, Ajan A, Hasegawa D, Okamoto I: *J Magn Magn Mater.* 2003, 260:261.

5.   Iwasaki S: *IEEE Trans Magn.* 1980, 16:71.

6.   Grundy PJ: *J Phys D Appl Phys.* 1998, 31:2975.

7.   Terris BD, Weller D, Folks L, Baglin JEE, Kellock AJ, Rothuizen H, Vettiger P: *J Appl Phys*. 2000, 87:7004.

8.   Hasegawa T, Pei W, Wang T, Fu Y, Washiya T, Saito H, Ishio S: *Acta Mater*. 2008, 56:1564.

9.   Chun DW, Kim SM, Kim GH, Jeung WY: *J Appl Phys*. 2009, 105:07B731.

10.   Hong MH, Hono K, Watanabe M: *J Appl Phys*. 1998, 84:4403.

11.   Zhang B, Soffa WA: *IEEE Trans Magn*. 1990, 26:1388.

12.   Sun S, Murray CB, Weller D, Folks L, Moser A:*Science*. 2000, 287:1989.

13.   Chen JS, Yingfan Xu, Wang JP:*J Appl Phys*. 2003, 93:1661.

14.   Zhu Yun, Cai JW:*Appl Phys Lett*. 2005, 87:1-032504.

15.   Qiu LJ, Ding J, Adeyeye AO, Yin JH, Chen JS, Goolaup S, Singh N:*IEEE Trans Magn*. 2007, 43:2157.

16.   Chen SC, Kuo PC, Kuo ST, Sun AC, Lie CT, Chou CY:*Mater Sci Eng*. 2003, B98:244.

17.   Cao J, Cai J, Liu Y, Yang Z, Wei F, Xia A, Han B, Bai J:*J Appl Phys*. 2006, 99:1-08F901.

18.   Xu Y, Chen JS, Wang JP:*Appl Phys Lett*. 2002, 80:3325.

19.   Murayama N, Soeya S, Takahashi Y, Futamoto M:*J Magn Magn Mater*. 2008, 320:3057.

20.   Chun DW, Kim SM, Jeung WYTo be published

21.   Breitling A, Goll D:*J Magn Magn Mater*. 2008, 320:1449.

22.   Zhong H, Tarrach G, Wu P, Drechsler A, Wei D, Yuan Jun:*Nanotechnology*. 2008, 19:095703.

23.   Lairson BM, Visokay MR, Sinclair R, Clemens BM:*Appl Phys Lett*. 1993, 62:639.

24.   Jaafar M, Sanz R, Vázquez M, Asenjo A, Jensen J, Hjort K, Flohrer S, McCord J, Schäfer R:*Phys Stat Sol (A)*. 2007, 204:1724.

25.   Kavita S, Raghavendra Reddy V, Gupta A, Amirthapandian S, Panigrahi BK:*Nucl Instrum Meth B*. 2006, 244:206.

26.   Jeong JR, Kim YS, Shin SC: *J Appl Phys*. 1999, 85:5762.

Chapter 5

A General Mathematical Programming Approach for Process Plant Layout

Michael C. Georgiadis[a, b], Gordian Schilling[a, c],
Guillermo E. Rotstein[a, d], and Sandro Macchietto[a]

[a]Centre for Process Systems Engineering, Imperial College of Science Technology and Medicine, London SW7 2BY, UK
[b]FORTH/CPERI, PO Box 361, Thessaloniki 57001, Greece
[c]Ciba Speciality Chemicals Inc., CH-1870 Monthey, Switzerland
[d]15 Babington Road, London NW4 4LD, UK

ABSTRACT

The generation of a good layout is an important stage in the design of a new plant or the retrofit of an existing facility. Layout decisions affect piping, electrics, instrumentation and therefore have a great impact on the total plant cost. Moreover, layout has a large impact on the safety, operability and maintainability of any chemical plant. This paper presents a general mathematical programming approach for addressing the problem of allocating items of equipment in a given two or three dimensional space. The problem is formulated as a mixed integer linear programming model where equipment of various sizes and geometries are taken into account. The objective function to be minimized accounts for the total transport, connection, land and floor construction cost. This optimization procedure results in the coordinates of each unit (location), the total piping length, and the land occupied. Three case studies are presented to illustrate the applicability of the proposed approach.

INTRODUCTION

Process plant layout is a creative task demanding significant engineering ingenuity and experience. It is concerned with the spatial arrangement of processing equipment and the interconnecting pipework. It relies on complex data and has a great impact on the plant operation. A good layout will account for safety requirements, environmental limitations, economy and maintenance accessibility and must achieve a good balance between these sometimes conflicting criteria.

Several methods have been proposed to solve the chemical plant layout problem. Most of them are based on heuristic rules and are restricted to two dimensional layout problems (Bradley & Nolan, 1985, Amorese, Cena & Mustacchi, 1991 and Fuchino, Itoh & Muraki, 1997). Optimization based techniques have also been presented. Layout considerations were taken into account in the design and scheduling of pipeless batch plants (Realff, Shah & Pantelides, 1996). For this type of plant the layout problem is of particular significance since the location of the processing stations determines the transfer times for the operating vessels. Penteado and Ciric (1996) presented an optimization formulation that can aid in the development of safe and economical layouts. The problem was solved as a relaxed mixed integer non linear programming (MINLP) and illustrated with a case study for the two dimensional layout.

The problem of facility layout and location is relatively old and has received over the years considerable attention in the field of operations research (Francis & White, 1974). Most of the published work has addressed the problem of job shops and assembly facilities and several formulations and algorithms have been proposed for the solution of these problems (Rosenblatt, 1979, Bozer, Meller & Erlebacher, 1994 and Fortenberry & Cox, 1985). Recently, Bazargan-Lari (1997) presented a mathematical model for layout design in cellular manufacturing problems. In this type of problem a poor shop floor layout can affect considerably the material movements and therefore the lead time of products to customers.

The application of operation research methods to design the layout of chemical or process plants is relatively new. Special attention must be given to the complex and unique characteristics of these plants. In this context, Jayakumar and Reklaitis (1994) proposed an approach for the single floor layout (involving grouping of equipment into different sections) establishing the analogy between distributing units on a single floor and the traditional graph partitioning problem. The same authors also considered the problem of partitioning of equipment in a multifloor chemical plant. A heuristic procedure was proposed which was complemented by a mathematical programming approach. The combination of both approaches was illustrated with the solutions of large scale problems (Jayakumar & Reklaitis, 1996). Suzuki, Fuchino, Muraki and Hayakawa (1991b) proposed a method for assigning each piece of equipment to any of the floors available in multipurpose plants. Since the distances between units were not calculated, approximate expressions were used for the transportation and floor construction cost. The same authors presented a method for multi-floor equipment arrangements in batch plants. Their approach assumes that the assigned floor for each equipment unit is known a-priori. The objective function is based only on weighted preferences for the allocation of equipment without economic considerations (Suzuki, Fuchino, Muraki & Hayakawa, 1991a).

Recently, we presented a mathematical formulation for addressing the two dimensional and three dimensional layout problem to the retrofit design of multipurpose batch plants (Georgiadis, Rotstein & Macchietto, 1997; Georgiadis & Macchietto, 1997). Trade-offs between capital and operating cost were captured so that the optimal number of floors may be determined.

One of the weaknesses of all the above methods is that equipment sizes and geometries are often not taken into account since units are allowed to be allocated to predetermined discrete locations. This may result in an overestimation or underestimation of the actual area and pipework required (for example if a small unit is allocated to a large area) and may also impose significant limitations during the later design stages.

Most of the previous work on the chemical plant layout using mathematical programming techniques has been mainly focused on the use of special formulations and address a part of the layout problem. A major deficiency is that equipment orientation is not considered. Also cost of land and floor construction cost are typically often not taken into account. Furthermore, many of the approaches are only restricted to two dimensional layout problems.

This paper extends our previous work (Georgiadis et al., 1997) and presents a general formulation and a mathematical programming solution for addressing the problem of allocating items of equipment in a given two or three dimensional space. Detailed cost factors are used to account for the cost of land, piping, transportation and floor construction cost. The approach takes into account the sizes and different rectangular geometric shapes of process equipment and allows for their proper orientation. Multilevel units (i.e. those which require space on more than one floor) are also considered. The proposed approach can be applied at the conceptual phase of plant design, when a plant flowsheet (giving number and type of each piece of equipment, and their connectivity) and the dimensions of the equipment are available. Three case studies are presented to illustrate the applicability of this approach to different type of problems.

PROBLEM STATEMENT

The problem addressed in this paper can be stated formally as follows:

Given:

- a set of process equipment units and their geometric shapes and sizes, indexed here as either i or j=1, . . . , n;
- the geometric shape of each piece of equipment as rectangles of various sizes;
- a set of available locations and their coordinates, indexed k=1, . . . , K;
- the material flowrates between connected units in kg s^{-1}.
- the purchase and installation cost of piping in \$ m^{-1}, floor construction cost and land cost in \$ m^{-2};
- minimum safety distance between units i and j when their proximity is considered as source of hazards.

Determine:

- the allocation of each unit to floors and the orientation of the unit;
- the area required on each floor for the three dimensional problems;
- the land required for the two dimensional problems.

Such that:

- the cost of layout including connection, transport, land and floor construction cost is minimized.

In this work the idea of uniform area division (UAD) is employed. The layout area is divided into a number of discrete equal grids (K) which are represented by unique x_k, y_k and z_k relative coordinates. For the sake of simplicity and with no loss of generality unit grids are assumed. In order to illustrate the representation used, consider the layout area shown in Fig. 1. For the first grid $x_1=1$, $y_1=1$, $z_1=1$. Similarly for grid 7 $x_7=2, y_1=2$, $z_1=1$. In case a second floor level is also included then all the corresponding grids to this level will have zk=2. This grid representation will aid to explain the problem constraints and objective function, in a better way, as shown latter. Each grid has the same side length, S_d, in the x and y coordinates. This assumption is necessary to state the rotation of equipment correctly. In order to model the problem accurately, the greatest common divisor of the size of all units must be used as the size of each grid. For example if there are four units of square shape with sizes 2, 4, 6 and 14 m² then the size of each grid will be 2 m².

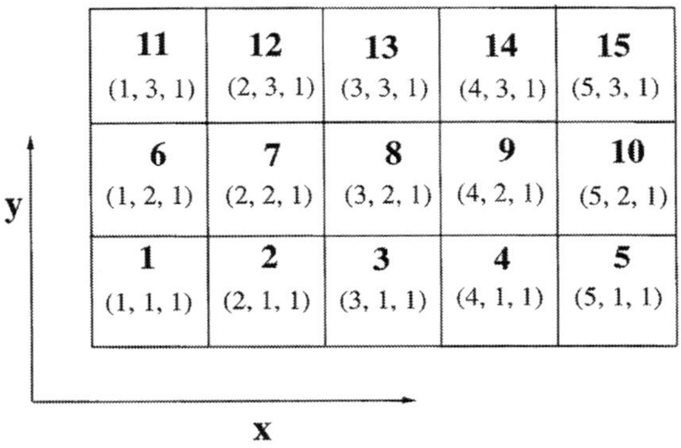

11	12	13	14	15
(1, 3, 1)	(2, 3, 1)	(3, 3, 1)	(4, 3, 1)	(5, 3, 1)
6	**7**	**8**	**9**	**10**
(1, 2, 1)	(2, 2, 1)	(3, 2, 1)	(4, 2, 1)	(5, 2, 1)
1	**2**	**3**	**4**	**5**
(1, 1, 1)	(2, 1, 1)	(3, 1, 1)	(4, 1, 1)	(5, 1, 1)

Figure 1: Grids representation in the layout area.

The shape of each unit i, is specified using a set of sections s which are connected to form a rectangle. A rectangular shape is considered more appropriate for industrial problems than circular shapes (some extra space is included to allow access for maintenance, inspection etc.). This discretization into sections is performed to enable the allocation of each section to a single grid. For the example with four units presented in the previous paragraph, the equipment of size 6 m² will be divided into three sections and the one of 14 m² into seven sections. The smallest unit includes only one section of size 2 m². If all the equipment has only one section then the formulation presented here reduces to that of our previous work (Georgiadis et al., 1997).

It should be emphasized here that in other approaches the layout problem is as a continuous nonlinear nonconvex optimization problem because the equations needed to prevent two units from occupying the same physical space are nonconvex. This results in a very difficult mathematical problem with local optimal solutions. The UAD proposed here avoids the problems of the nonlinearities and nonconvexities by transforming the problem into a mixed integer linear problem which can be solved using standard commercial MILP solvers. Obviously, if the discretization used in the model is too coarse, the UAD will lead to a suboptimal solution.

MATHEMATICAL FORMULATION

For each equipment one must specify which of the sections defining the shape are used for connectivity purposes. For example if a heat exchanger has a rectangular shape, (3×1) sections, XXX with section numbering 123 one can define the connection sections as 1 and 3 (inlet and/or outlet). The definition of the connection sections is based on the physical form of the equipment and it is not a degree of freedom available for the optimizer. The optimization procedure will determine the optimal allocation of the connection sections but each equipment has to follow a well defined rectangular shape with all the relative sections allocated close to each other.

The direction matrix between two connected units represents now the connectivity between two sections.

$d_{isjt} = 1$,

> if flow is from section s of unit i, to section t of unit j

$d_{isjt} = 0$,   otherwise

Basic Variables

The variables representing distances between sections are based on the grid representation and denote relative values expressed as integer numbers. For example, the horizontal distance between grid 1 and 3 in Fig. 1 is 2 (i.e. two times the width of a grid). In order to calculate the actual distance, the horizontal distance variables (in the x–y level) are multiplied by the grid side S_d. Similarly, the vertical distance variables are multiplied by the floor height, F_h. For all the connections between unit sections (i.e. disjt=1), we introduce a set of variables representing the relative distance (in integer values) in the x-, y- and z-coordinates:

R^x_{isjt}: relative distance in x-coordinates between section s of unit i and section t of unit j, if unit i is to the right of unit j.

L^x_{isjt}: relative distance in x-coordinates between section s of unit i and section t of unit j, if unit i is to the left of unit j.

R^y_{isjt}: relative distance in y-coordinates between section s of unit i and section t of unit j, if unit i is to the right of unit j.

L^y_{isjt}: relative distance in y-coordinate between section s of unit i and section t of unit j, if unit i is to the left of unit j.

U_{isjt}: relative vertical distance between section s of unit i and section t of unit j, if unit i is at a higher floor than unit j.

D_{isjt}: relative vertical distance between section s of unit i and section t of unit j, if unit i is at a lower floor than unit j.

In addition, a binary assignment variable yisk is defined as:

$y_{isk} = 1,$

   if section s of unit i is allocated to site area k

$y_{isk} = 0,$ otherwise

Note that all variables apart from y_{isk} are treated as continuous throughout the optimization.

Objective Function

The objective function is to minimize the total cost by considering:

- the upward and horizontal transportation cost;
- the cost of piping;
- the floor construction cost (three dimensional problems);
- the land cost (two dimensional problems);

The allocation of units to different floors requires the use of cost factors which differ from those in the two dimensional case. The piping cost for the movement of materials to higher floors cannot be neglected and must be included in the objective function. In the case of a downward flow there is often no pumping cost due to gravity, but the connection cost ($ m^{-1}) still exists. Finally, where two connected units are allocated to the same floor, a (low) pumping and a connection costs are taken into account. This objective function provides a good estimation of financial costs. It compares favorable with other works, where arbitrary cost values are typically used especially where distances cannot be calculated explicitly (Suzuki et al., 1991a).

For each connection between section s of unit i and section t of unit j three different yearly cost factors are proposed:

- Upward cost, CF^{up}isjt: there is a high cost associated with transport of materials to higher floors. This cost is a function of the flow-rate between the connected process units and the height differences. This function is given approximately by Coulson and Richardson (1985):

$$CF^{up}_{isjt} = \text{Energy\_unit\_cost} \cdot \text{Flow} \cdot g \quad \$ \cdot m^{-1} \cdot \text{year}$$

- The above cost must be multiplied by the corresponding variables which represent the height difference. The gravity constant g is taken as 9.81 m s^{-2}, the flow must be given in kg s^{-1} and the unit transport cost in $ kW^{-1} h^{-1} (energy cost). The above expression is multiplied by 1.2 to account for other factors such as the bends in the pipes, the roughness of the pipe surface etc. The final expression is:

$$CF^{up}_{isjt} = \beta^{up} \cdot \text{Flow} \quad \$ \cdot m^{-1} \cdot \text{year}$$

- where β^{up} is a case dependent constant which for the purpose of our analysis is taken equal to 1515.
- Downward flow, CF^{down}isjt: in this case there is often no pumping cost since the flow is gravity driven. However, there is a connection cost which is a function of the vertical distance between the connected units.

- Horizontal cost, $CF2^{hor}isjt$: this term represents the cost to transport the material on the same floor. A low transport cost together with a connection cost is taken into account. The pumping cost is a function of flow-rate and distance. It is estimated as 10% of the upward pumping cost and is given as follows (Coulson and Richardson, 1985):

$$CF^{down}_{isjt} = \beta^{hor} \cdot Flow \quad \$ \cdot m^{-1} \cdot year$$

- where β^{hor} is a case dependent constant equal to 151.5 for the problems considered here. This cost coefficient must be multiplied by the horizontal distance.

In addition to the cost of transferring material between two connected units, the piping cost must be also considered (annualised through a capital charge factor).

Summarizing, the following input data are required:

- $CF^{up}isjt$: up-ward flow cost of unit i section s to unit j section t in $ m⁻¹ year.
- $CF^{down}isjt$: down-ward flow cost of unit i section s to unit j section t in $ m⁻¹ year.
- $CF^{hor}isjt$: horizontal flow cost of unit i section s to unit j section t in $ m⁻¹ year.
- $CCisjt$: connection cost between unit i section s and unit j section t in $ m⁻¹.
- Cost of land: for two dimensional layout problems in $ m⁻².
- FCC: floor construction cost for the three dimensional layout problems in $ m⁻².
- FA: area per floor for the three dimensional layout problems in m².
- CCF: capital charge factor (Douglas, 1988).
- F_h: floor height for the three dimensional problems in m.
- S_d: length of each grid side in m.
- NF: number of floors.

Thus the objective for the three dimensional layout problem becomes:

$$\min OF = \sum_{(ts),(jt)|d_{isjt}=1} [(CF_{isjt}^{down} + CC_{isjt}) \cdot U_{isjt} \cdot F_h$$
$$+ (CF_{isjt}^{up} + CC_{isjt}) \cdot D_{isjt} \cdot F_h$$
$$+ (CF_{isjt}^{hor} + CC_{isjt})(R_{isjt}^x + L_{isjt}^x + R_{isjt}^y + L_{isjt}^y)$$
$$\times \cdot S_d] + CCF \cdot FCC \cdot FA \cdot NF$$

$$+ CCF \cdot LA \cdot LCC \tag{1}$$

where LA is the land area in m² which is calculated explicitly as shown in Section 3.7

If a two dimensional layout is considered then the term representing the floor construction cost is neglected. In principle, the number of floors, NF, can be a variable and special constraints accounting for the calculation of the optimal number of floors must be added, as in our previous work (Georgiadis & Macchietto, 1997). However, it is beyond of the scope of this paper to consider the problem of the optimal number of floors. Alternatively, and given that the number of floors is usually small, the problem presented here can be solved parametrically in terms of NF for the case of multilevel problems.

Assignment Constraints

Each unit (and the corresponding sections) must be allocated to some of the given K locations. On the other hand, each location may be occupied at most by one unit. The following constraints are thus imposed:

$$\sum_{k=1}^{K} y_{isk} = 1 \ \forall \ i, s \tag{2}$$

$$\sum_{i=1}^{n} \sum_{s}^{N_{seci}} y_{isk} \leq 1 \ \forall \ k \in [1, K] \tag{3}$$

where $N_{sec}i$ is the number of sections for unit i.

Distance Constraints

The distance between connected units is taken to be rectangular which is more appropriate in industrial problems than straight-line distances. The following constraints calculate the relative distance between two connected units:

$$R_{isjt}^x - L_{isjt}^x = \sum_{k=1}^{K} x_k \cdot (y_{isk} - y_{jtk}) \forall [(is),(jt)] \mid d_{isjt} = 1 \tag{4}$$

$$R_{isjt}^y - L_{isjt}^y = \sum_{k=1}^{K} y_k \cdot (y_{isk} - y_{jtk}) \forall [(is),(jt)] \mid d_{isjt} = 1 \tag{5}$$

$$U_{isjt} - D_{isjt} = \sum_{k=1}^{K} z_k \cdot (y_{isk} - y_{jtk}) \forall [(is),(jt)] \mid d_{isjt} = 1 \tag{6}$$

where x_k, y_k, z_k are the relative coordinates of grid K (see Fig. 1). The distance between two connected units should always be no less than one (i.e. one grid size). The following constraint is thus added:

$$R_{isjt}^x - L_{isjt}^x + R_{isjt}^y + L_{isjt}^y + U_{isjt} + D_{isjt} \geq 1 \ \forall [(is),(jt)]$$

$$|d_{isjt} = 1 \tag{7}$$

The constraints (7) are redundant, however, they are added to obtain a tighter formulation. Note that all the above constraints are only defined for the connection sections of the two units.

It should be emphasised that the problem being solved involves minimisation of a distance related objective function thus minimising the $R_{isjt}^x + R_{isjt}^y + L_{isjt}^x + L_{isjt}^y + U_{isjt} + D_{isjt}$ summation. Consequently, only one variable at most of each pair—i.e. (R_{isjt}^x, L_{isjt}^x), (R_{isjt}^y, L_{isjt}^y), (U_{isjt}, D_{isjt})—is guaranteed to be non-zero at the optimal solution of each linear programming (LP) problem during the branch-and-bound procedure used for the MILP solution.

Equipment Shape Constraints: Single Level

As mentioned before the geometric shape of each equipment is described by several sections s each the size of one grid in the x–y plane. First we assume that each equipment is allocated to one floor. Special constraints must be added to describe the geometric shape of each unit. Each unit is taken to have a rectangular footprint which is more appropriate for industrial problems than circular footprints. The orientation of the equipment can be modified to obtain an optimal process layout.

Basic Linear Shape

In order to take into account different equipment sizes we first consider simple forms which provide the building blocks to construct more complex ones. The simplest unit is described by just one section. Obviously, no constraints are required to describe this shape. The next simplest equipment shape consists of two sections XX with section numbering **12**. To enforce this shape in the layout area we must impose a set of constraints. First we introduce the following constraints:

$$\sum_{k=1}^{K} (x_k + y_k) \cdot (y_{i1k} - y_{i2k}) = 1 - 2o_i$$

$$(8a)$$

where o_i is an integer variable representing the orientation of the equipment, parallel to either the x or yaxis, and excluding the unfeasible diagonal section allocation. For example, in Fig. 1 the allocation of such a unit to grid one (section 2) and seven (section 1) is prohibited since the above constraint is violated. The meaning o_i is better described graphically in Fig. 2. The above constraint does not avoid geometries with $\Delta x=2$ and $\Delta y=-1$, for example allocation to grid three (section 1) and six (section 2) (seeFig. 1). Thus, we need to ensure that the relative distance in the x- and y-coordinates does not exceed one, so as to exclude these prohibited allocations. The following constraints are then imposed:

$$-1 \le \sum_{k=1}^{K} x_k \cdot (y_{i1k} - y_{i2k}) \le 1 \qquad |x^{\dim} > 2$$

(8b)

$$-1 \le \sum_{k=1}^{K} y_k \cdot (y_{i1k} - y_{i2k}) \le 1 \qquad |x^{\dim} > 2$$

(8c)

where $x^{\dim}$ and $y^{\dim}$ is the total number of grids in one floor for the x and y coordinate, respectively. For example in Fig. 1, $x^{\dim}=5$ and $y^{\dim}=3$. Note that these constraints are only written if $x^{\dim}$ and $y^{\dim}$ are greater than two. In that case it is possible to have the absolute difference in the x or y coordinate greater than one. In any other case these constraints are satisfied automatically by (8a).

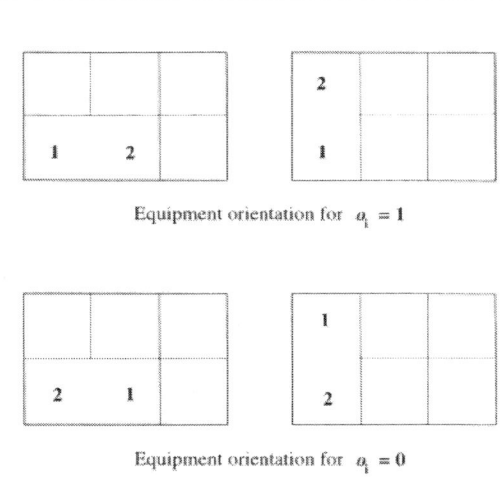

Figure 2: Equipment orientation in the x–y level.

Each section of the equipment has to be allocated on the same floor, i.e. $\Delta z=0$:

$$\sum_{k=1}^{K} z_k \cdot (y_{i1k} - y_{i2k}) = 0 \qquad |z^{\dim} > 1$$

(8d)

where $z^{\dim}$ is the number of floors. The above constraints will be only imposed if more than one floor is considered.

Now using this basic shape we can easily derive the constraints for related shapes, the most common of which are presented as follows.

The third case is XXX with section numbering 123. The constraints (8a–d) need to be extended by the following set to link section 3 correctly with section 2:

$$\sum_{k=1}^{K} x_k \cdot (2y_{i2k} - y_{i1k} - y_{i3k}) = 0$$

(9a)

$$\sum_{k=1}^{K} y_k \cdot (2y_{i2k} - y_{i1k} - y_{i3k}) = 0$$

(9b)

Since the distance between section 1 and 2, due to (8b and c), is equal to the side of one grid, the above constraints describe the unique geometric shape of the equipment. Note that the orientation has been already determined by (8a). In addition we need to ensure that section 3 is on the same level as sections 1 and 2:

$$\sum_{k=1}^{K} z_k \cdot (y_{i2k} - y_{i3k}) = 0 \qquad |z^{\dim} > 1$$

(9c)

The fourth case is X X X X with section numbering 1 2 3 4. The constraints (8a–d) and (9a–c) need to be extended in order to link section 4 correctly with section 3. The following constraints are thus imposed:

$$\sum_{k=1}^{K} x_k \cdot (2y_{i3k} - y_{i2k} - y_{i4k}) = 0$$

(10a)

$$\sum_{k=1}^{K} y_k \cdot (2y_{i3k} - y_{i2k} - y_{i4k}) = 0$$

(10b)

Note that sections 1 and 3 are forced on the same floor by constraint (9c) and similar constraints are needed for sections 3/4:

$$\sum_{k=1}^{K} z_k \cdot (y_{i3k} - y_{i4k}) = 0 \quad \left|z^{\text{dim}} > 1\right.$$

(10c)

It should be noted here that every additional section for the linear shape requires three constraints each of the form (9a, b and c).

Two Dimensional Shapes:

Using the previously presented base-forms it is possible to construct two dimensional shapes by using a new binary variable li. This variable links the basic shapes in order to achieve proper orientation of the equipment. The idea is to describe the two dimensional shape as a link of more than one simple one dimensional shapes.

Consider the square shape:

X   X

X   X

with the section numbering

3   4

1   2

First one describes the geometrical shape of base sections 1/2 and 3/4 by applying the previously presented type of constraints (8a–d). A unique binary variable o_i is used for each unit i forcing an identical orientation of these sections. Next the two basic shapes are linked as follows:

$$\sum_{k=1}^{K} (x_k + y_k) \cdot (y_{i3k} - y_{i1k}) = 1 - 2l_i$$

(11a)

$$-1 \le \sum_{k=1}^{K} x_k \cdot (y_{i3k} - y_{i1k}) \le 1 \quad \left|x^{\text{dim}} > 2\right.$$

(11b)

$$-1 \leq \sum_{k=1}^{K} y_k \cdot (y_{i3k} - y_{i1k}) \leq 1 \quad |x^{\dim} > 2$$

(11c)

$$\sum_{k=1}^{K} (x_k + y_k) \cdot (y_{i4k} - y_{i2k}) = 1 - 2l_i$$

(11d)

$$-1 \leq \sum_{k=1}^{K} x_k \cdot (y_{i4k} - y_{i2k}) \leq 1 \quad |x^{\dim} > 2$$

(11e)

$$-1 \leq \sum_{k=1}^{K} y_k \cdot (y_{i4k} - y_{i2k}) \leq 1 \quad |y^{\dim} > 2$$

(11f)

where the new binary variable l_i is defined similar to o_i, to allow proper orientation of sections 1/3 and 2/4. The above constraints ensure that the equipment will be allocated in such a way so as to follow its specific geometric shape.

In addition, all sections should be allocated to the same floor and this is guaranteed by the following constraints between sections 1/3 and 2/4:

$$\sum_{k=1}^{K} z_k \cdot (y_{i3k} - y_{i1k}) = 0 \quad |z^{\dim} > 2$$

(11g)

$$\sum_{k=1}^{K} z_k \cdot (y_{i4k} - y_{i2k}) \leq 1 \quad |z^{\dim} > 2$$

(11h)

Note that constraints (11a–f) cannot prevent the mirror geometry

$X_1 \quad X_2$

$X_3 \quad X_4$

of the original shape

$X_3 \quad X_4$

$X_1 \quad X_2$

For this purpose, the following constraints are added in the model:

$$\sum_{k=1}^{K} x_k(y_{i1k} - y_{i2k}) = 1 - l_i - o_i$$

(12a)

It can be easily verified that all mirror geometries violate the above constraints. Similarly, other shapes can be modelled, such as

$$\begin{array}{ccc} X & X & X \\ X & X & X \end{array}$$

with the section numbering

$$\begin{array}{ccc} 4 & 5 & 6 \\ 1 & 2 & 3 \end{array}$$

First constraints (8a–d) and (9a–c) are used to describe the geometrical form of the two base shapes given by sections 1/2/3 and sections 4/5/6. Then these two forms are linked using similar constraints to (11a–h) for the linkage between sections 1/4 and 5/2. Finally, in order to avoid mirror geometries, equivalent to (12a) constraints are added each involving oi and li and two adjacent sections.

A similar approach can be followed to describe a 3×3 square given as:

$$\begin{array}{ccc} X & X & X \\ X & X & X \\ X & X & X \end{array}$$

with the section numbering

7 8 9
4 5 6
1 2 3

This is modelled by considering constraints (8a–d) and (9a–c) to describe the geometrical form of the three base shapes given by sections 1/2/3, 4/5/6 and 7/8/9. Then these three base shapes are linked together using constraints analog to (11a–h) for linking any two pairs of sections 1/4 to 3/6 and similarly of sections 4/7 tc 6/9. Constraints equivalent to (12a) are also included to avoid mirror geometries.

Constraints similar to (11g and h) are imposed to ensure that all the equipment sections are allocated to the same floor.

Equipment Constraints: Multi-Level

The model presented in the previous section considers the case where each equipment item is not allowed to occupy more than one floor. However, in many practical application an equipment (e.g. distillation column) must be allocated space on more than one floor.

If an equipment uses more than one floor, it is straightforward to link the previously described shapes in thex–y coordinates together with the shape description in the z-coordinate. Since we do not allow rotation in the z-coordinates we know from the section enumeration, which sections are going to be allocated to higher floors and which have the same x–y coordinates.

The simplest equipment consists of two sections one above the other,

X

X

with numbering

2

1

For this equipment the following constraints describe its geometric shape:

$$\sum_{k=1}^{K} x_k \cdot (y_{i2k} - y_{i1k}) = 0$$

(13a)

$$\sum_{k=1}^{K} y_k \cdot (y_{i2k} - y_{i1k}) = 0$$

(13b)

$$\sum_{k=1}^{K} z_k \cdot (y_{i2k} - y_{i1k}) = 1$$

(13c)

where (13a and b) ensure that the x and y coordinates of section 1 and 2 are identical, while (13b) that section 2 is on top of section 1. If the equipment has the shape

X   X

X   X

with the section numbering

3   4

1   2

where sections 1 and 2 are to be allocated to a lower floor than sections 3 and 4 then in addition to the basic equations (8a–d) for sections 1/2 and 3/4 the following constraints must be imposed:

$$\sum_{k=1}^{K} x_k \cdot (y_{i3k} - y_{i1k}) = 0$$

(14a)

$$\sum_{k=1}^{K} x_k \cdot (y_{i4k} - y_{i2k}) = 0$$

(14b)

$$\sum_{k=1}^{K} y_k \cdot (y_{i3k} - y_{i1k}) = 0$$

(14c)

$$\sum_{k=1}^{K} y_k \cdot (y_{i4k} - y_{i2k}) = 0$$

(14d)

where constraints (14a) and (14b) ensure that the x coordinates of sections 1 and 3 and sections 2 and 4 are identical while constraints (14c) and (14d) that the y-coordinates of all sections are also identical.

Finally the following constraints are added to ensure that section 3 and 4 are above 1 and 2.

$$\sum_{k=1}^{K} z_k \cdot (y_{i3k} - y_{i1k}) = 1$$

(15a)

$$\sum_{k=1}^{K} z_k \cdot (y_{i4k} - y_{i2k}) = 1$$

(15b)

If the equipment consists of three sections one above the other,

X
X
X

with the numbering

3
2
1

For this equipment further to the constraints (13a–c) the

following are also imposed to connect section 3 with the other two:

$$\sum_{k=1}^{K} x_k \cdot (y_{i3k} - y_{i2k}) = 0 \tag{16a}$$

$$\sum_{k=1}^{K} y_k \cdot (y_{i3k} - y_{i2k}) = 0 \tag{16b}$$

$$\sum_{k=1}^{K} z_k \cdot (y_{i3k} - y_{i2k}) = 1 \tag{16c}$$

In a similar fashion one can describe more complex two dimensional and three dimensional shapes.

Calculation of the Land Area

In many countries, land is expensive and a compact plant layout is preferable to minimize its cost. Here, it is assumed that the required area is defined by the minimum rectangle which includes all the allocated units. This approach is restricted to the two dimensional layout problems. For example, consider the two dimensional plant layout in Fig. 3 where the area is divided into 15 equal grids each with an area of 100 m². If process units are allocated to grids 1, 2, 8, 9, 11, 12 and 14, it is clear that the required area is determined by a rectangle with length w and width d. The extra area denoted by grids 5, 10 and 15 is not really required.

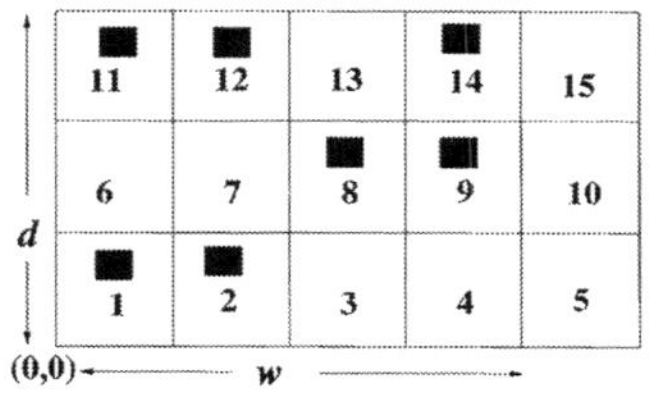

Figure 3: Determination of the required land in the two dimensional problems.

The idea is to determine the optimal area of land based on the unique grid that describes the above rectangle. In order to do so, to each grid in the two dimensional space we assign an area ARk defined by the area of the rectangle with respect to the origin of the axis. For example (see Fig. 3) for grid 1 this area will be 100 m², for grid 8, 600 m² (i.e. six internal grids) and for grid 14, 1200 m². It is clear that grid 14 is the one that will determine the required area. Now a new integer variable is defined for each grid as follows:

$q_k = 1$,

   if grid k determines the rectangular area required

$q_k = 0$,   otherwise

The following set of constraints is added to the mathematical model.

$$\sum_k q_k = 1 \tag{17}$$

$$\sum_k x_k \cdot q_k = w \tag{18}$$

$$\sum_k y_k \cdot q_k = d \tag{19}$$

$$\sum_i \sum_s y_{isk} \cdot x_k \leq w \quad \forall k \tag{20}$$

$$\sum_i \sum_s y_{isk} \cdot y_k \leq d \quad \forall k \tag{21}$$

Finally the required land LA is given by the following equation:

$$LA = \sum_k AR_k \cdot q_k \tag{22}$$

The above formulation can easily be extended to nor-equal grids.

Note that the cost of land area, LA, can also be included in three dimensional layout problems. In this case LA will be equal to the floor area, FA, and thus optimally decided by the problem. Clearly, no modification is required for constraints (17–22).

Degeneracy Reduction Constraints

The proposed formulation presents many degenerate solutions: given any solution, one may generate others with exactly the same value of the objective function. The source of this problem is that the formulation relies on relative distances between the units. For instance, equivalent solutions may often be obtained from the previous one by shifting its allocation by one column or row in a discrete area. The existence of degenerate solutions may have adverse effects on the efficiency of the search procedure. To avoid it without any loss of optimality, one can introduce the following constraints, similar to those proposed in Georgiadis et al. (1997).

$$\sum_i \sum_i \sum_{k \in L_m(k)} y_{isk} \le \sum_i \sum_s \sum_{k \in L_{m-1}(k)} y_{isk} \cdot MNU$$

$$(23)$$

where $L_m(k)$ and $L_{m-1}(k)$ are two neighbouring subsets of areas (rows or columns in the two dimensional layout area or floor). MNU is the maximum number of units which can be allocated in each subset of areas considered. For example according to Fig. 4 the above constraints must be applied for L_2 (grids 5, 6, 7, 8) and L_1 (grids 1, 2, 3, 4) where MNU is 4. Similarly, the same constraints will be also applied for L_5 (grids 2, 6, 10) and L_4 (grids 1, 5, 9). In that case MNU is 3. It is clear that for the simple example considered in Fig. 4 the above constraints will enforce the creation of the solution denoted by circles instead of its equivalent denoted by 'crosses', thus resulting in a more compact layout.

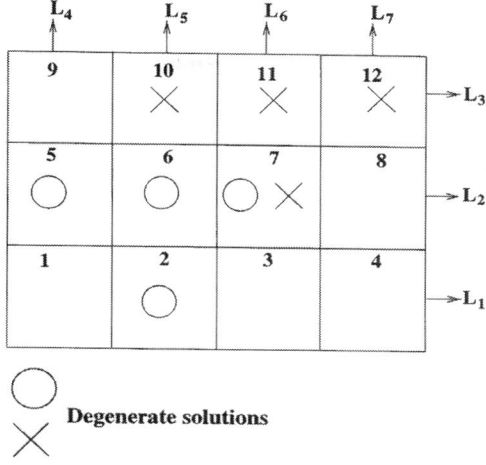

Figure 4: Degeneracies in the optimal plant layout problem.

It should be added that the above constraint (23) is not required if the objective function includes the cost of land area since the latter will also exclude the generation of the solution represented by 'crosses' as illustrated in Fig. 4.

Comments

The objective function and all constraints are linear including binary variables. Thus, the formulation presented in this section is a MILP problem which can be solved using standard MILP solvers. A number of equipment shapes has been modelled and proper orientation in the x–y level is allowed to achieve the optimal layout. Equipment orientation on the z-coordinate is not considered here. Although the above general formulation considers the three dimensional layout, two dimensional problems are a special sub-case where the continuous variables related with the z dimension are zero.

It is clear that any equipment size and rectangular shape can be straightforward modelled for single or multi level units by imposing constraints similar to the above and/or suitable combinations of them. For the sake of simplicity and without loss of generality it is considered that there is no need to model more equipment shapes here.

The equipment division into sections greatly affects the size of the above MILP. If there is a big difference between the size of the smallest and largest equipment then a fine discretization is required and the problem will be large even for small process flowsheets (since the largest equipment will be divided into many sections). One way to avoid this difficulty is to use aggregate equipment modules. A module would include a group of units which have similar operating characteristic or high degree of interconnection. For example a distillation column and its associated condenser, reboiler, pumps and reflux drum can form a module. Once the choice of modules has been made, and equipment in each module has been specified (including relative position), the plant layout problem takes the form of deciding the relative position of modules. In practice, further simplification to the problem can be achieved due to safety or operational constraint which may prohibit the allocation of some units to special areas or impose their allocation to fixed areas. In that case the corresponding integer variables yisk will be set either to zero (exclude allocation) or to one (impose allocation). In the special case where two units must be allocated on the same floor (due to operating considerations) their vertical distance in the z direction should be set to zero. Furthermore, if two units must be allocated close to each other then the total distance in the x–y level between the two connection sections will be set to the size of one grid. Finally, constraints to impose a minimum distances between units which are considered source of hazard can be easily imposed.

EXAMPLES

The above MILP formulation has been implemented in a C++, object-oriented environment and solved using the cplex V4.0 mixed integer linear programming package (CPLEX, 1994). Once the decision of the layout area discretization into grids and equipment into sections has been made, the user can easily define in an input file the number of grids in the x and y direction, the number of floors (for three dimensional problems), the size of each grid and height of floors and the geometric shape of each process unit. Each process unit is described by a vector the element of which are the number of sections in all directions. For example a unit described as [2, 2, 1] consists of two sections along the x-coordinate, two other along the y-coordinate and will be allocated to a single floor (square shape). The grids coordinates are then automatically generated. Constraints such as allocation of process units to specific areas or floors can be easily imposed. The connections between units, the corresponding connection sections and cost data are also easily defined. A brief description of the solving procedure is given as follows:

- read input file into the data structure;
- setup the MILP with the current data;
- solve the MILP;
- read the solution of the MILP solver and produce output file;

The output file includes the exact allocation of each unit, the distances between connected units and the value of the objective function. For all the case studies presented below an optimality margin of 1% has been used.

Example 1

Consider the small instant coffee process taken from Jayakumar and Reklaitis (1996) and shown in Fig. 5. The equipment sizes in terms of rectangular footprints and the cost data are given in Table 1 and Table 2, respectively. The geometrical shape of each process unit and their sections are shown in Fig. 6. The size of each grid is defined to be 10 m^2 and each equipment is divided into a number of sections each of size one grid. Note that these data are only used for illustrative purposes. Here the objective function considers only the transport and piping cost in order to illustrate the effect of the number of floors on the operating cost. It is beyond the scope of this paper to consider trade-offs between capital and operating cost. The reader can refer to our previous work where such trade-offs are illustrated resulting in the optimal number of floors required (Georgiadis & Macchietto, 1997). The following cases are considered:

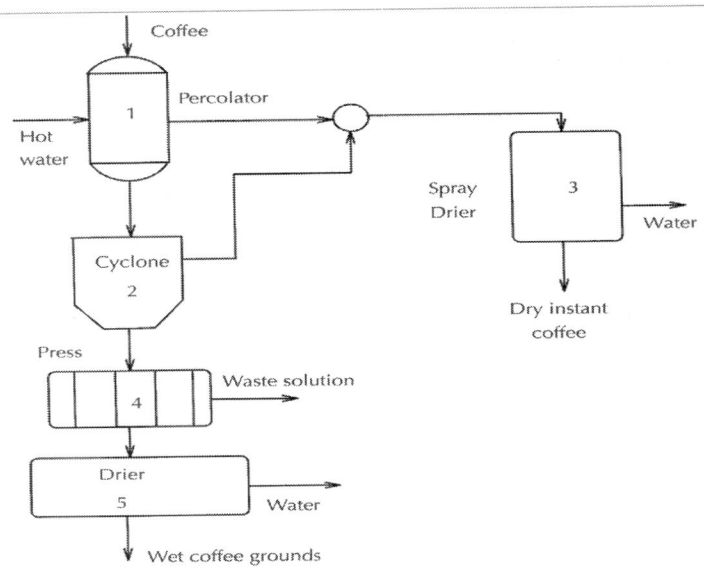

Figure 5: Flowsheet of instant coffee process.

Table 1: Equipment sizes for example 1

Equipment	Size (m²)
1	50
2	10
3	50
4	40
5	30

Table 2: Cost data for example 1

Stream	CF^{up}_{ij} ($ m^{-1} year^{-1})	CF^{hor}_{ij} ($ m^{-1} year^{-1})	CCO_{ij} ($ m^{-1})
(1, 2)	25 250	2520	600
(1, 3)	37 820	3720	800
(2, 3)	6300	630	350
(2, 4)	18 820	1880	400
(4, 5)	14 190	1410	500

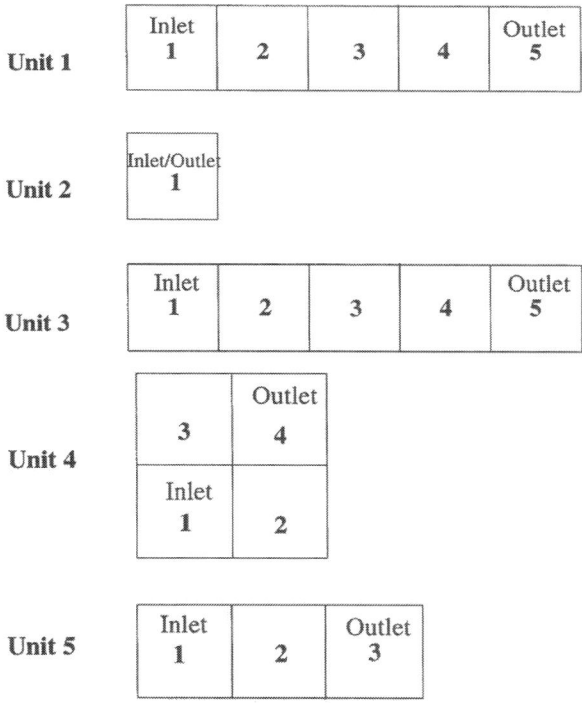

Figure 6: Processing units shapes for example 1.

Two Dimensional Layout

The layout area is divided into 36 grids each with a size of ˉ0 m². The objective function comprises land, transport and piping cost. Since the land cost is uncertain and may differ significantly from one country to another (or even for different areas within the same country) we consider three sub cases:

- Zero land cost: in that case only transport and piping costs are considered. The optimal layout is shown in Fig. 7. Although the transport cost is minimized the obtained layout is not compact. The total cost is 48 200 $ year^{-1} and the layout area 300 m².

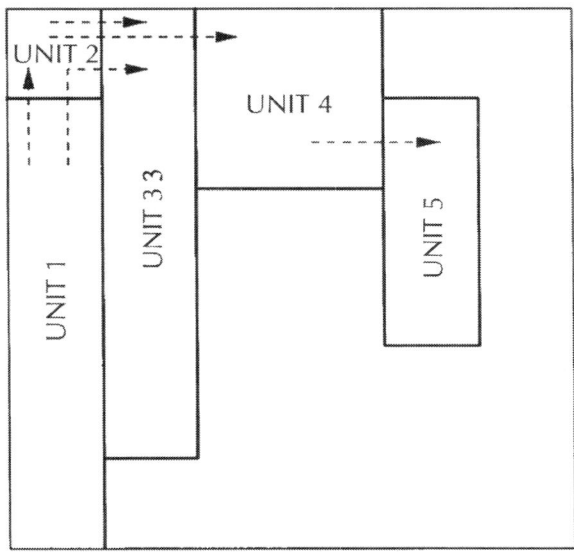

Figure 7: Optimal two dimensional layout for example 1-case 1.

- The land cost coefficient is taken to be 300 $ m^{-2}: here the optimal layout is depicted in Fig. 8. As we can see the units are allocated in such a way that the horizontal transport cost is minimized and the obtained layout is more compact than the previous case due to the land cost. The total layout cost is 118 900 $ year^{-1} and the layout area 240 m² (20% less than in the previous scenario).

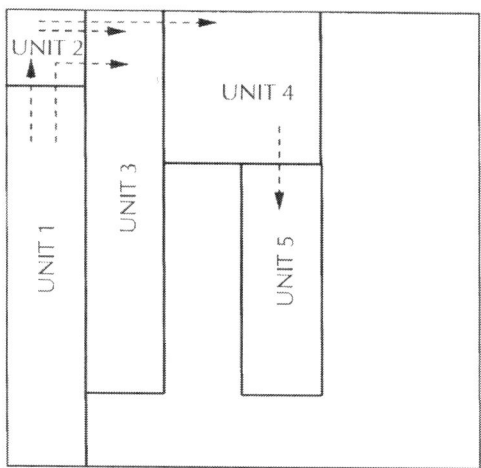

Figure 8: Optimal two dimensional layout for example 1-case 2.

- The land cost coefficient takes the value of 1000 $ m^{-2}: A different layout is obtained and shown in Fig. 9. It is clear that a more compact layout is obtained, minimizing the cost of land (and increasing the pumping cost) which represents a significant factor in the objective function. The area required is now 200 m^2 and the layout cost 278 400 $ $year^{-1}$.

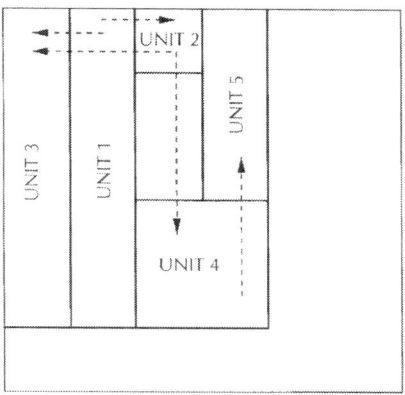

Figure 9: Optimal two dimensional layout for example 1-case 3.

Two Floor Levels

The layout area is divided into 30 grids (15 grids on each floor). There are five grids in the x and three grids in the y dimension. The height of each floor is taken to be 5 m. The optimal process layout is depicted in Fig. 10. We observe that the units are allocated in such a way that a compact layout is obtained eliminating the high upward transport cost. An important information from the resulting layout is that only 100 m^2 are required for the second floor while the first floor is totally utilized. The cost of layout is 41 300 \$ year^{-1} and there are three horizontal transport flows (between units 1/2, 2/3 and 4/5). Note that unit 1 is exactly allocated on top of unit 3.

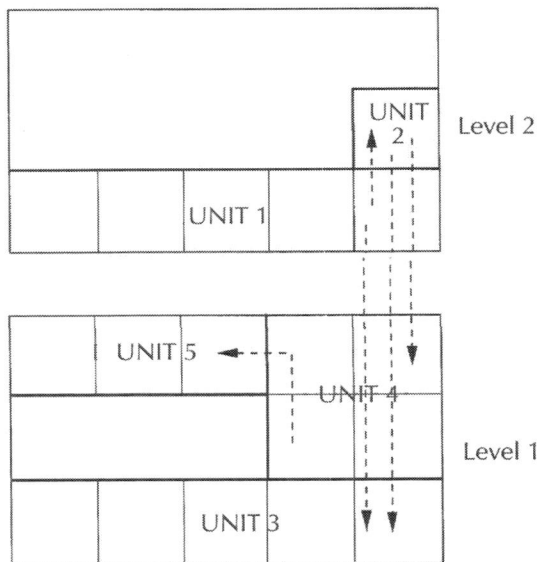

Figure 10: Optimal layout on two levels for example 1.

Three Floor Levels

Here the layout area is divided into 45 grids (15 grids on each floor). The optimal layout is depicted in Fig. 11. The process units are allocated one below the other in a compact way, so that as many as possible flows are aided by the gravity. The cost of layout is 37 500 $ year⁻¹ and it is lower compared with the two level problem since there are two horizontal transport costs (between units 1/2 and 2/3) and all the others are zero due to gravity flows. It is interesting to note that unit 5 is vertically oriented with respect to the xaxis.

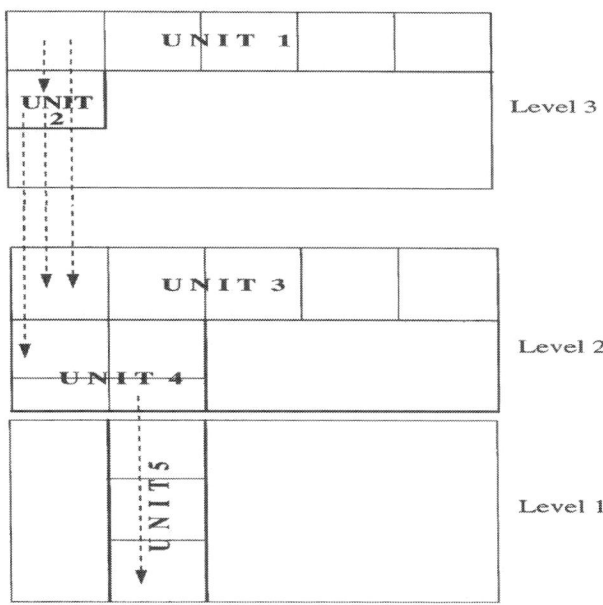

Figure 11: Optimal layout on three levels for example 1.

Four Floor Levels

Now there are 60 grids (15 grids on each floor). The optimal layout is shown in Fig. 12. The same trend as in the previous cases holds. The process units are allocated so as to make the high upward cost to be zero. There is only one horizontal transport cost between units 4/5. This results in a slightly lower operating cost of 36 200 $ year⁻¹ comparing with the three level case. The resulting layout indicates also the required area for each floor. Thus, level one requires 100 m², level two, 50 m², level three, 10 m² and level four 50 m².

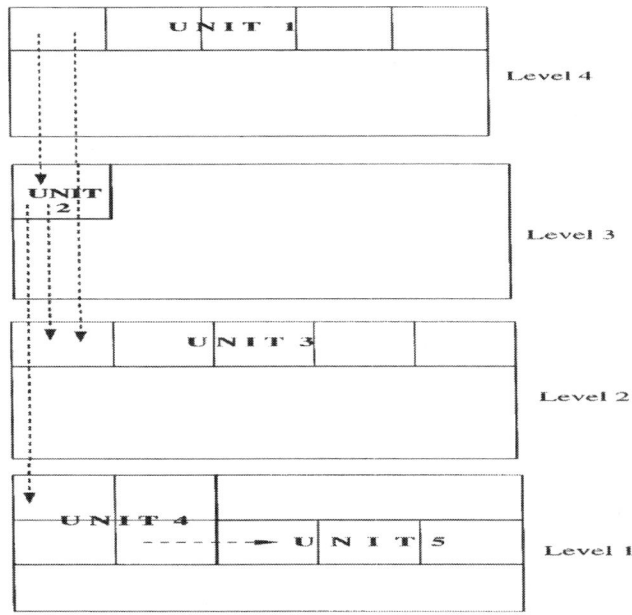

Figure 12: Optimal layout on four levels for example 1.

Three Floor Levels-Multilevel Equipment

Here a multilevel equipment layout is considered. Unit 5 is assumed to occupy three floors with one section above the other. Similarly unit 4 is required to occupy two floors with sections 1 and 2 above sections 3 and 4. The optimal layout is shown in Fig. 13. It is clear that unit 4 is allocated to all floors and unit 5 to the first and second floor. Again, the upward transport cost is minimized to zero but there is a substantial horizontal pumping cost compared with the other cases which results in a higher layout cost (45 600 $ year^{-1}). This is due to limitations imposed by the multilevel nature of the two units. It should be emphasized that the decision of multilevel equipment is based only on its physical shape and not on economic considerations. For example a tall batch distillation column has to be allocated in more than one floor.

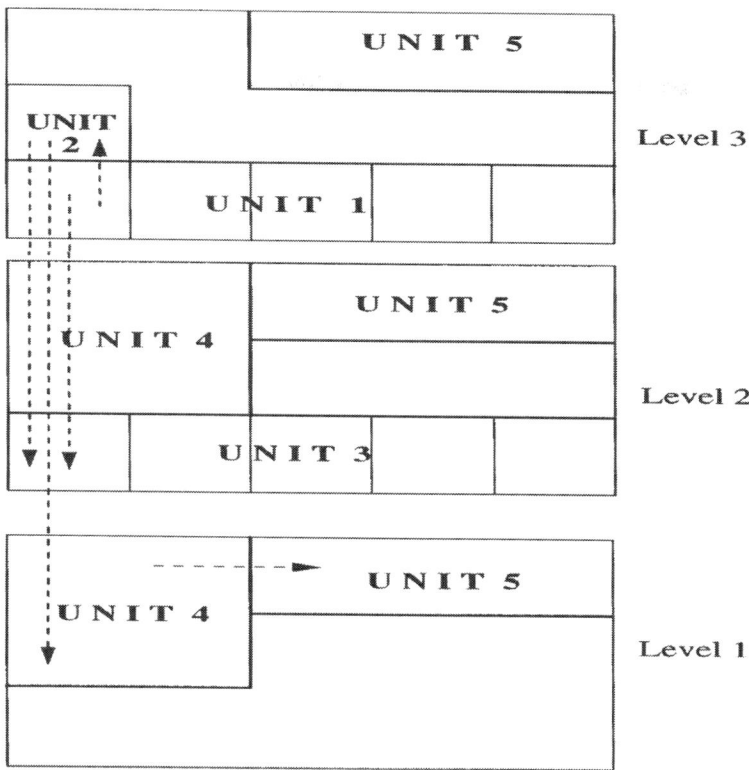

Figure 13: Optimal layout on three levels with multilevel units for example.

Example 2

The process flowsheet shown in Fig. 14 is used as the second case study. It has been derived from the example considered by Suzuki et al., 1991a and Suzuki et al., 1991b. The geometric shape of each process unit is depicted in Fig. 15. Multilevel equipment are not considered. Each floor has a height of 5 m and is divided into 18 grids. The size of each grid as determined by the smallest equipment is 10 m² (seeTable 3). Cost data concerning the transport and connection cost are given in Table 4.

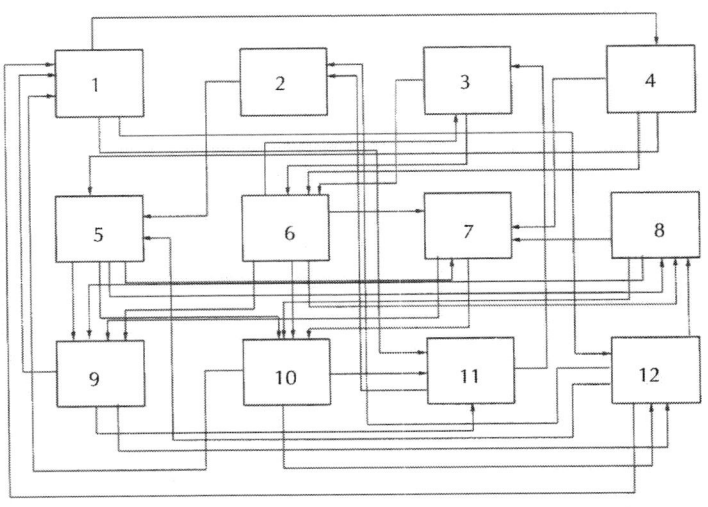

Figure 14: Process flowsheet for example 2.

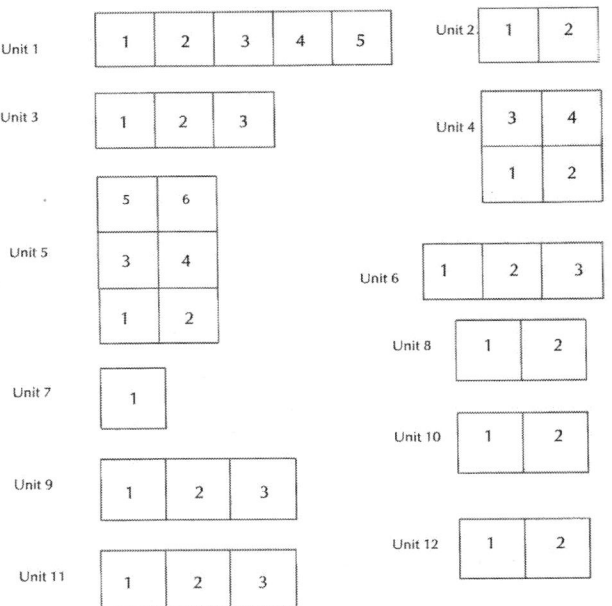

Figure 15: Processing units shapes for example 2.

Table 3: Equipment sizes for example 2

Equipment	Size (m²)
1	50
2	20
3	30
4	40
5	60
6	30
7	10
8	20
9	30
10	20
11	30
12	20

Table 4: Cost data for example 2

CF^{up}ij	Stream $ m^{-1} year^{-1}	CF^{hor}ij $ m^{-1} year^{-1}	CCOij $ m^{-1}	Stream	CF^{up}ij $ m^{-1} year^{-1}	CF^{hor}ij $ m^{-1} year^{-1}	CCOij $ m^{-1}
(1,4)	2000	200	200	(7,4)	2000	200	200
(1,11)	2000	200	200	(7,9)	2000	200	200
(1,12)	2000	200	100	(7,10)	2000	200	250
(2,5)	4000	400	400	(8,5)	4000	400	400
(2,11)	2000	200	260	(8,7)	2000	200	300
(2,12)	2000	200	340	(8,9)	2000	200	230
(3,5)	4000	400	170	(8,10)	2000	200	300
(3,6)	4000	400	310	(9,1)	2000	200	200
(3,9)	2000	200	260	(9,6)	4000	400	300
(3,10)	2000	200	200	(9,7)	2000	200	370
(4,5)	4000	400	110	(9,1)	2000	200	160

(4,6)	4000	400	230	(9,12)	2000	200	100
(4,7)	2000	200	290	(10,1)	2000	200	400
(4,8)	2000	200	370	(10,5)	4000	400	200
(5,7)	4000	400	400	(10,11)	2000	200	190
(5,8)	4000	400	440	(11,1)	2000	200	300
(5,9)	4000	400	230	(11,2)	2000	200	300
(5,10)	4000	400	150	(11,3)	2000	200	300
(5,3)	4000	400	100	(11,7)	2000	200	240
(6,7)	4000	400	360	(11,9)	2000	200	230
(6,8)	4000	400	220	(12,1)	2000	200	400
(6,9)	4000	400	500	(12,2)	2000	200	260
(6,10)	4000	400	310	(12,5)	2000	200	190
(12,8)	2000	200	130	(12,10)	2000	200	220

The optimal layout is depicted in Fig. 16. We can see that a compact layout is obtained in an attempt to minimize the high upward transport cost. It is interesting to note again that the units have been oriented properly according to their specific geometric shape. The optimal layout corresponds to an yearly cost of 480 000 $ year^{-1}.

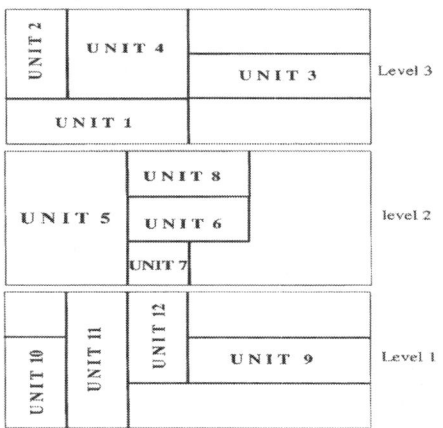

Figure 16: Plant layout for example 2.

Example 3

Consider the case study from Barbosa-Povoa and Macchietto (1994) shown in Fig. 17. It is an industrial multipurpose batch plant, with a high degree of connectivity and 18 processing units. The size and shape of each process unit are shown in Table 5 and Fig. 18, respectively. Cost data with respect to the transport and connection cost are given in Table 6. A five floor process layout is considered. Each floor is divided in ten equal grids with size 10 m². There are five grids in the x coordinate and two in the y coordinate. Each floor is taken to have a height of 5 m. Units still, v9 and v5 are to be allocated in two floors.

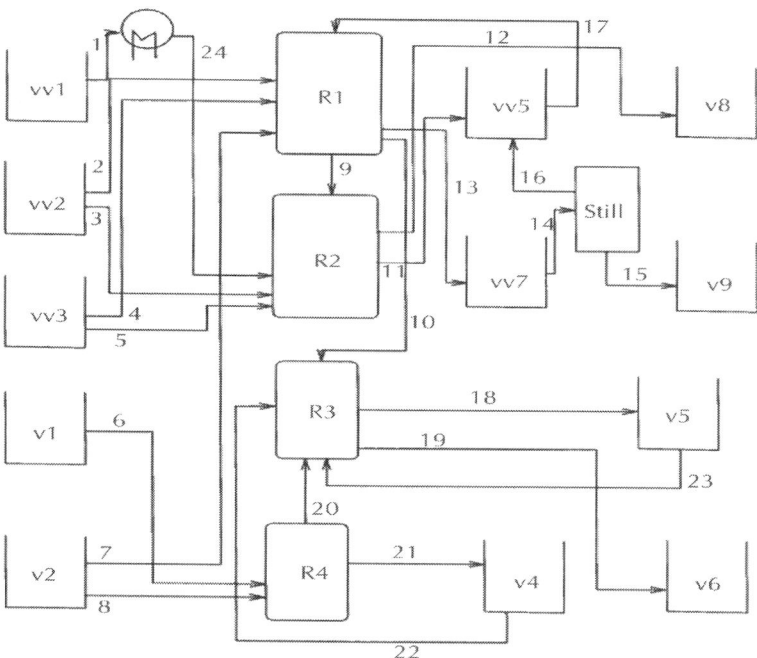

Figure 17: Flowsheet of example 3.

Table 5: Equipment sizes for example 3

Equipment	Size (m²)	Equipment	Size (m²)
vv1	20	vv5	30
vv2	20	vv7	30
vv3	30	v4	30
v1	20	still	20
v2	30	v8	30
R1	40	v9	20
R2	40	v5	20
R3	30	v6	30
R4	20	Heater	10

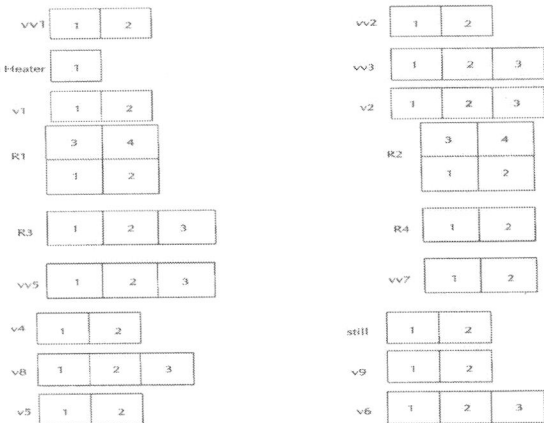

Figure 18: Processing units shapes for example 3.

Table 6: Cost data for example 3

Stream	CF^{up}ij ($ m^{-1}year^{-1})	CF^{hor}ij ($ m^{-1}year^{-1})	CCOij ($ m^{-1})	Stream	CF^{up}ij ($ m^{-1}year^{-1})	CF^{hor}ij ($ m^{-1}year^{-1})	CCOij ($ m^{-1})
(vv1, heater)	3250	325	200	(vv$_3$, R2)	2540	254	230
(Heater, R1)	3680	368	240	(v$_1$, R4)	2690	269	160

(Vv2, R2)	3540	354	230	(v2, R1)	3680	368	250
(vv3, R1)	3680	368	400	(v2, R4)	2690	269	160
(R1, R2)	3540	354	230	(R1, R3)	2800	280	170
(R2, vv5)	5300	530	270	(R2, v8)	5300	530	270
(R1, vv7)	5500	550	280	(vv7, still)	5500	550	300
(Still, vv5)	2760	276	170	(still, v9)	2760	276	170
(vv5, R1)	8100	810	300	(R3, vv5)	4100	410	250
(R3, v6)	4100	410	250	(R4, R3)	2800	280	175
(R4, v4)	2690	269	170	(v4, R3)	2800	280	175
(vv5, R3)	1200	120	140				

The optimal layout is shown in Fig. 19. We can see that a compact layout is obtained minimizing the upward pumping cost. Unit still is allocated to the second and third floor and units v9 and v5 to the first and second floor. The cost of the layout is 111 000 $ year^{-1} comprising transport and connection costs.

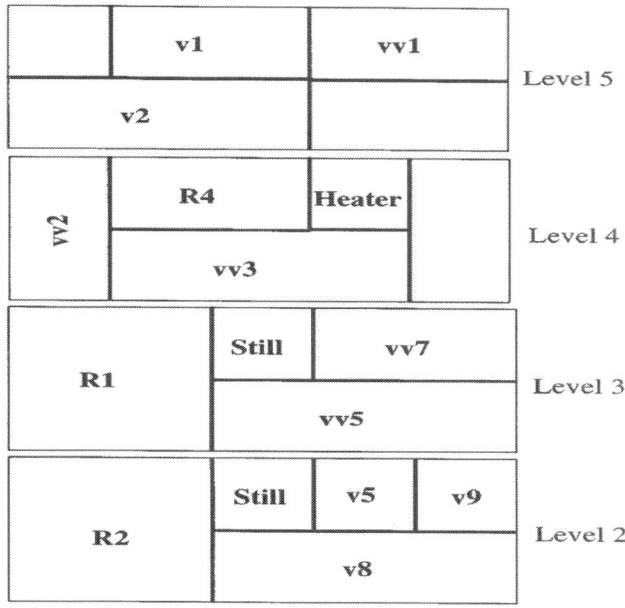

Figure 19: Plant layout for example 3.

CONCLUSIONS

An MILP formulation is developed for the two dimensional and three dimensional process plant layout problems based on a uniform area discretization approach. It has been tested for the solution of problems with up to 20 units. The objective function represents the total transport, piping, land and floor construction cost using detailed cost models and incorporating the aspect of direction dependent flow costs. The effect of land cost on the plant layout has been investigated with a small process example. Here, high land cost may results in more compact layouts. A salient feature of the formulation is that equipment sizes and geometries based on rectangular shapes are considered when solving the problem of finding the optimal allocation. Both single level and multilevel equipment geometries are considered. Orientation of the equipment in the two dimensional level is allowed in order to obtain optimal layouts. Different types of constraints, for example due to safety considerations (e.g. minimum distance between units) can be easily incorporated. A way to handle degenerate solutions is also presented. The effectiveness of the presented approach is demonstrated with the attainment of global optimality for three case studies with up to 18 units.

It should be emphasized, however, that for problems with more than 20 units and for cases where many different equipment sizes and geometries exist the presented approach generates a formulation with a potentially very large number of binary variables. This may lead to the solution of an MILP problem requiring high computational costs. In that case the problem can be considerably simplified if the optimal partitioning of units to floors is given (for example using the approach of Jayakumar & Reklaitis (1996)). Once the optimal partitioning has been decided, the problem can take the form of deciding about the optimal exact allocation of process units (incorporating their sizes and geometries) using the approach presented in this paper. Thus, even for problems of industrial nature a good initial layout can be obtained during the conceptual design stage.

ACKNOWLEDGMENTS

The authors would like to thank the unknown reviewers for the useful comments and suggestions. Financial support by EPSRC is also gratefully acknowledged.

REFERENCES

1. Amorese, L., Cena, V., & Mustacchi, C. (1991). A heuristic for the compact location of process components. Chemical Engineering and Science, 32, 119–124.

2. Barbosa-Povoa, A. P., & Macchietto, S. (1994). Detailed design of multipurpose batch plants. Computers and Chemical Engineering, 18, 1013–1042.

3. Bazargan-Lari, M. (1997). Inter-cell layout design in a cellular manufacturing environment—a case study. Flexible automation and intelligent manufacturing (pp. 504–516). Begel House.

4. Bozer, Y., Meller, R., & Erlebacher, S. (1994). An improved type layout algorithm for single and multiple-floor facilities. Management Science, 40, 918–932.

5. Bradley, C. W., & Nolan, P. F. (1985). Criteria for plant separation, distances and location. Industrial Chemistry and Engineering Symposium, 93, 247–262.

6. Coulson, J. M., & Richardson, J. F. (1985). Chemical Engineering (3). Oxford: Pergamon.

7. CPLEX, (1994). Optimization using the CPLEX barrier and mixed integer sol6ers. CPLEX, Houston, TX.

8. Douglas, J. M. (1988). Conceptual design of chemical processes (1). New York: McGraw-Hill.

9. Fortenberry, B. J. C., & Cox, J. F. (1985). Multiple Criteria approach to the facilities layout problem. Int. J. Prod. Res., 23, 773–782.

10. Francis, R. L., & White, J. A. (1974). Facility Layout and Location (1). Englewood Cliffs, NJ: Prentice-Hall.

11. Fuchino, T., Itoh T., & Muraki, M. (1997). Arrangement of process equipment modules under the consideration of plant safety. Journal of Chemical Engineering of Japan (in press).

12. Georgiadis, M. C., & Macchietto, S. (1997). Optimal layout of process plants: a novel approach. Computers and Chemical Engineering, S2, S337–S342.

13. Georgiadis, M. C., Rotstein, G. E., & Macchietto, S. (1997). Optimal layout design in multipurpose batch plants. Industrial Engineering and Chemical Research, 36, 4852–4863.

14. Jayakumar, S., & Reklaitis, G. V. (1994). Chemical plant layout via graph partitioning: (I) single level. Computers and Chemical Engineering, 14, 441–458.

15. Jayakumar, S., & Reklaitis, G. V. (1996). Chemical plant layout via graph partitioning: (II) multiple levels. Computers and Chemical Engineering, 20, 563–578.

16. Penteado, A., & Ciric, A. R. (1996). An MINLP approach for safe process plant layout. Industrial Engineeering and Chemical Research, 35, 1354–1361.

17. Realff, M. J., Shah, N., & Pantelides, C. C. (1996). Simultaneous design, layout and scheduling of pipeless batch plants. Computers and Chemical Engineering, 20, 869–883.

18. Rosenblatt, M. J. (1979). The facilities layout problem: a multi-goal approach. International Journal of Production Research, 17, 323– 332.

19. Suzuki, A., Fuchino, T., Muraki, M., & Hayakawa, T. (1991a).

20. Equipment arrangement for batch plants in multifloor buildings with integer programming. Journal of Chemical Engineering of Japan, 24, 737–742.

21. Suzuki, A., Fuchino, T., Muraki, M., & Hayakawa, T. (1991b). Method of determining the floor for siting of each equipment unit of a multipurpose batch plant (in English). Kagaku Kogaku Ronbunshu, 17, 1110.

An Efficient Method for Optimal Design of Large-scale Integrated Chemical Production Sites with Endogenous Uncertainty

Sebastian Terrazas-Moreno[a], Ignacio E. Grossmann[a], John M. Wassick[b], Scott J. Bury[b], and Naoko Akiya[b]

[a]Department of Chemical Engineering, Carnegie Mellon University, Pittsburgh, PA 15213, United States

[b]The Dow Chemical Company, Midland, MI 48674, United States

ABSTRACT

Integrated sites are tightly interconnected networks of large-scale chemical processes. Given the large-scale network structure of these sites, disruptions in any of its nodes, or individual chemical processes, can propagate and disrupt the operation of the whole network. Random process failures that reduce or shut down production capacity are among the most common disruptions. The impact of such disruptive events can be mitigated by adding parallel units and/or intermediate storage. In this paper, we address the design of large-scale, integrated sites considering random process failures. In a previous work (Terrazas-Moreno et al., 2010), we proposed a novel mixed-integer linear programming (MILP) model to maximize the average production capacity of an integrated site while minimizing the required capital investment. The present work deals with the solution of large-scale problem instances for which a strategy is proposed that consists of two elements. On one hand, we use Benders decomposition to overcome the combinatorial complexity of the MILP model. On the other hand, we exploit discrete-rate simulation tools to obtain a relevant reduced sample of failure scenarios or states. We first illustrate this strategy in a small example. Next, we address an industrial case study where we use a detailed simulation model to assess the quality of the design obtained from the MILP model.

INTRODUCTION

The optimal design and operation of integrated production networks is a current and future opportunity in the chemical process industry. For instance, The Dow Chemical Company owns Texas Operations, an integrated site that manufactures 21% of the company's products sold globally (Wassick, 2009). BASF's site in Ludwigshafen is another example of a large integrated production system with over 200 production plants (BASF, 2010). Both of these sites began as smaller manufacturing facilities and grew in capacity and complexity over many decades. In contrast with the gradual integration of these heritage sites, recent strategic initiatives require the grassroots design of very large integrated process networks. The joint venture between Saudi Aramco and The Dow Chemical Company to construct and operate a world-scale chemical and plastic production complex in Saudi Arabia is an example of such an initiative (Dow, 2007).

These integrated sites feature different interconnected process networks. A failure event that reduces the production rate of any of the processes can propagate throughout the network. Some events are planned, for example, preventive maintenance of major plant components; the remaining failure events occur at random, requiring corrective maintenance. The focus of this work is on the second type of events, namely, failure modes that decrease production capacity and that occur at random times with random repair durations. The industrial significance of this problem is illustrated in a paper by Miller, Owens, and Deans (2006) from The Dow Chemical Company. These authors explain the benefit of designing reliability into manufacturing systems and illustrate the scope of the involvement of Reliability–Availability–Maintainability (RAM) teams during the design of large-scale manufacturing systems. A note on terminology; availability is the ratio of uptime to total time or the fraction of time the unit is producing product, reliability is the probability of a unit or piece of equipment being in an up state at a particular time.

To study these stochastic failures, computer simulations are commonly used to test the effect of design alternatives in the availability or effective capacity of integrated systems. This approach is applicable to an integrated chemical facility. As expected for an integrated system, the increasing number of design degrees of freedom and the increasing number of stochastic inputs quickly increases the difficulty and time to search the design space, requiring more computing and modeling resources. In addition, the successful ranking and selection of options is often challenging to do cleanly when the search space is large. We believe that mathematical programming techniques can be used in conjunction with process simulation tools to provide an efficient tool for improving the design of integrated sites by significantly reducing the design space. This new space can be then thoroughly explored and validated via simulation. Additional opportunities for applying optimization to the solution of the design, planning, and scheduling of integrated sites can be found in Wassick (2009). In our previous work (Terrazas-Moreno, Grossmann, Wassick, & Bury, 2010), we proposed a mixed-integer linear programming (MILP) model for the optimal design of integrated sites subject to random failures and random supply and demand. The optimization criteria were the maximization of the probability of meeting customer demands and the minimization of capital investment. To be a practical and useful tool for design teams, this challenging model must be solved efficiently. A literature review of optimization approaches for related problems, which includes the works by Davies and Swartz (2008), Pistikopoulos, Thomaidis, Melin, and Ierapetritou (1996), Pistikopoulos, Vassiliados, Arvela, and Papageorgiou (2001), and Straub and Grossmann, 1990 and Straub and Grossmann, 1993, can be found in that paper.

In the present paper, we propose a novel algorithm based on Benders decomposition (BD) (Benders, 1962) to solve large-scale instances of the MILP model mentioned above. There is rich literature on the application of BD to the design of process systems under uncertainty. Most of these applications model uncertainty using a stochastic programming (SP) representation and apply variations of the BD algorithm. The standard decomposition technique is referred to as L-shaped decomposition (Van Slyke & Wets, 1969) in the stochastic programming literature. Straub and Grossmann (1993) proposed a nonlinear programming (NLP) model for maximizing the feasible operating region of a network with uncertain process parameters and used Generalized Benders Decomposition (GBD) (Geoffrion, 1972) to solve this problem. A similar approach was proposed by Pistikopoulos (1995) and applied by Ierapetritou, Acevedo, and Pistikopoulos (1996) and Acevedo and Pistikopoulos (1998) as a general algorithmic technique for solving a class of problems defined as process design and operations under uncertainty. More recently, Liu, Fan, and Ordonez (2009) addressed the design of reliable transportation networks subject to unpredictable natural disasters using Generalized Benders Decomposition (GBD). The work by Santoso, Ahmed, Goetschalckx, and Shapiro (2005) is related to the model and algorithm we present in this paper, although they deal with exogenous uncertainties as opposed to endogenous uncertainties in our case. The paper by Santoso et al. (2005) proposes a two-stage stochastic programming (SP) model to optimize the design of a supply chain network under uncertainty. The authors consider in their MILP model uncertainty in parameters such as processing cost, raw material supply, fin shed product demand, and processing capacity of manufacturing facilities. These uncertain parameters are discretized in order to build scenarios with different combinations of parameter values. The resulting number of scenarios can be huge for realistic problem instances. The paper proposed an algorithm based on BD where the 1st stage network design variables are considered complicating. An interesting feature of this paper is that Benders decomposition is enhanced with convergence accelerating techniques based on three ideas. The first is adding constraints besides the usual dual cuts to the master problem that can be derived as strengthened dual cuts or from constraints expressed in terms of variables in the master problem that were redundant and not included in the full space model. The second idea is a heuristic for finding good feasible solutions. The third is a trust region algorithm that prevents the master problem from oscillating wildly in the first iterations.

All of the implementations of BD and GBD mentioned above correspond to SP problems with *exogenous*uncertainties. That is, the stochastic process is independent of design decisions (Jonsbraten, Wets, & Woodruff, 1998). If we exclude exogenous uncertainties like demand and raw material supply, then the source of remaining uncertainties are random process failures, which are dependent on design decisions. This implicitly assumes that when a unit is "up" that it operates at the design rates. The number and selection of parallel processing units in the network are not known *a priori*, i.e., they are decision variables. Since the only processes where random failures can occur are those that are selected from the superstructure, the realizations of uncertainties are also a function of the decision variables. This type of uncertainty, defined as *endogenous*, significantly increases the complexity of the problem as well as the computational resources required for solving it. This paper proposes a novel implementation of Benders decomposition for two-stage stochastic programming (SP) problems with endogenous uncertainties. The technique we propose partly overcomes the combinatorial explosion in problem size that occurs with non-anticipativity constraints required in this type of problems (Jonsbraten, Wets, & Woodruff, 1998).

In the following sections, we present the problem of optimal design of integrated sites and its representation as a two-stage MILP stochastic programming problem. Some important properties of the problem are given, and the decomposition approach that exploits these properties is presented. Finally, we test the methodology with two numerical examples, one of them being an industrial case study. As part of the solution of this case study, we analyze a scenario reduction technique, and we report the results of simulating the operation of the Pareto-optimal designs with a commercial simulation tool, ExtendSim®(Imagine That Inc., 2010).

PROBLEM STATEMENT

In this section, we describe the problem of optimal design of highly available integrated sites.

The following list of given data, degrees of freedom, optimization criteria, and assumptions is reproduced with minor changes from Terrazas-Moreno et al. (2010).

Given are:

- The superstructure of an integrated site with allowable parallel production units in each plant and intermediate storage tanks.
- A set of materials that the plants consume and produce.
- Mass balance coefficients for all units in the superstructure.
- The maximum capacity that can be assigned to each unit in the integrated site.
- Maximum supply of raw materials and maximum demand of finished products.
- Number of failure modes, production rate loss as a result of each failure, and the time between failure (TBF) and

time to repair (TTR) per failure mode, either as probability distributions or mean values (MTBF and MTTR, respectively).

- A cost function that relates design decisions with capital investment.
- The problem is to determine:
- The number of production units for each plant.
- The capacity of each unit.
- Sizes of intermediate storage between plants.
- For each state, material flows between plants and rate of accumulation or depletion of material in storage.
- The objective is to determine the set of Pareto-optimal solutions that:
- Maximize the average production rate at which the integrated site supplies chemicals to an external market.
- Minimize the capital investment.

An interesting aspect of our approach is the development of a set of discrete states that correspond to all possible combinations of failure modes in the integrated site and a set of frequency and duration equations that allows us to calculate the mean residence time mrt and frequency of encounters fr of all possible states in the system, based only on the knowledge of the $MTTR$ and $MTTF$ of the units in the superstructure (Billinton & Allan, 1992).

The following assumptions and simplifications are made:

- Random failures are independent events.
- The cost function is represented by piece-wise linear approximations.
- The production units in the plants are dedicated single product continuous processes.

Multiple failures in real integrated systems can be causally related. If this is the case, our first assumption will result in inaccurate statistical information of the discrete states described above. In practice, this is rare except at the equipment component level, which is at a level of detail finer than the failures considered in this work. Consequently, most reliability simulation is carried out assuming independent failure modes. In this sense, the independence assumption made in this work will not add any more inaccuracy than what is regularly used in industrial simulations. Statistically uncorrelated events allow us to determine discrete state probabilities, residence times, frequency, etc. using standard frequency and duration calculations (Billinton & Allan, 1992). Using a piecewise linear approximation of the objective function preserves linearity in the problem. The number of piecewise functions can be increased or decreased depending on the level of precision required in cost calculations. Finally, the third assumption limits the applicability of the approach to systems that can be modeled as continuous dedicated plants. Multiproduct plants would require scheduling considerations and a system of multiple storage tanks.

OVERVIEW OF SOLUTION STRATEGY

The solution approach we use in this paper integrates simulation and optimization tools. The simulation is built as a discrete rate model in Extend Sim software (Imagine That Inc., 2010). In a discrete rate model, flow rates and tank levels are updated only when needed, by an internal linear programming (LP) solver at each discrete event. In our problem, events are failures and repairs that change the rate of flow of materials among plants and changes in tank level status (e.g. full, empty, high, low). The optimization step involves a mixed-integer stochastic programming representation of the integrated site and exploits state-of-the-art computational technology for solving mixed-integer programs. The main idea of stochastic programming (SP) is to build a set of scenarios that consists of discrete realizations of uncertainties in the problem, and then optimizing the expected value of the objective criterion over all possible scenarios. In our problem, the scenarios correspond to the discrete failure states.

Combining the two technologies (simulation and optimization) exploits their complementary advantages. Discrete rate simulation is able to represent the operation of the integrated site in great detail. In fact, these types of models have been used at The Dow Chemical Company to simulate real manufacturing systems and validate them against actual operating data. Each simulation run, however, requires that design variables be fixed. Searching for an optimal design requires enumeration techniques that are time consuming and provide no guarantee of finding the optimal solution. By building and solving mixed-integer linear programming (MILP) models, one can find a guaranteed optimal solution using state-of-the-art MILP solvers such as CPLEX (ILOG, 2011). MILP models, however, are equation-oriented: the objective function and operational constraints have to be expressed in algebraic terms. In order to do so, the modeler is required to make certain simplifications of real systems. We use optimization to find a set of Pareto-optimal designs using an algebraic model of the system and simulation to evaluate in more detail the performance of these candidate designs. If the number of scenarios (discrete states) used in the optimization model is impractically large, as is the case in when there is a large number of possible failure modes, we also use the simulation model to generate a sample of the most representative scenarios. Finally, we determine some sensitivity analysis derived from the optimal solution to provide guidelines for fine-tuning the design using the simulation model. Fig. 1 summarizes the solution process described above.

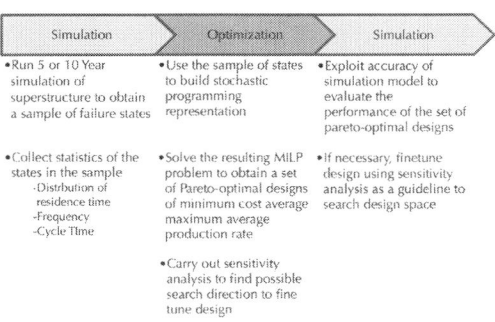

Figure 1: Summary of solution strategy using simulation and optimization tools.

The following sections describe in more detail each of the elements in the simulation and optimization solution approach. First, we develop the MILP model of the integrated sites and explain how the resulting formulation corresponds to a two-stage stochastic programming (SP) problem with endogenous uncertainties. The SP formulation relies on constructing a set of failure scenarios that can be impractically large for the problem at hand. To overcome the computational complexity involved in solving large instances of the problem, we propose a novel decomposition algorithm based on Benders decomposition. Next, we describe the discrete-rate model used to simulate the integrated site. Finally, we present two case studies: an academic example and an industrial case study. The first example is meant to illustrate the results of the optimization approach. Since the number of failure modes is small, we can enumerate all discrete failure states, and there is no need for a simulation model to obtain a representative sample of states. We use this small example to introduce the methodology for sensitivity analysis. Since the industrial case study is significantly larger, it requires the simulation–optimization–simulation sequence depicted inFig. 1. We use this example to explain the methodology for constructing a sample of failure states.

An important note to keep in mind is that in the SP representation, the discrete states that correspond to combinations of failure modes are called scenarios. For our purposes, the terms states and scenarios are equivalent, but the latter is used in some of the remaining sections in order to be consistent with the SP literature.

MIXED-INTEGER LINEAR PROGRAMMING MODEL

The starting point for the mathematical model of the integrated site is a superstructure that corresponds to a network of processes and storage tanks. Each of the nodes in the network has the structure shown in Fig. 2. The variables are described in the nomenclature section in Appendix A.

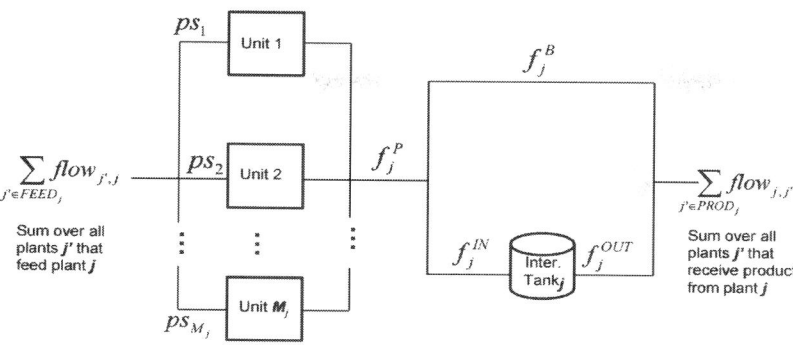

Figure 2: Building block for plant j in the integrated site.

In previous work (Terrazas-Moreno et al., 2010), a mixed-integer linear programming (MILP) formulation for solving this problem was proposed in which exogenous uncertainties in supply and demand were also included. The most important features of the formulation are: (i) a superstructure that contains all potential parallel units and storage tanks in all plants in the process network, (ii) a state-space representation of the integrated site where each discrete state corresponds to a combination of simultaneous failure modes, and (iii) a model of intermediate storage, based on the concept of random walks (Heyman & Sobel, 1982), to determine the average and variance of the levels of material in the storage tanks as a function of the network design. The transitions between states (random process failures and repairs) follow the behavior of a continuous time Markov chain (Heyman & Sobel, 1982). This approach allows the calculation of the mean and variance of the time spent in each state, the frequency of visits to each of them, and the probabilities of finding the integrated site in any state using statistical information from historical reliability data of existing processes that resemble the ones postulated in the superstructure.

All of the above elements are integrated in an MILP formulation that maximizes the expected stochastic flexibility [E(SF)] of finished products and minimizes the capital investment required by the network design (Terrazas-Moreno et al., 2010). The bi-criterion optimization problem is solved using the ε-constrained method (Ehrgott, 2005). The degrees of freedom are the selection of units from the superstructure, the size and location of intermediate storage tanks, and the production capacity of the plants. An important difference between our previous paper and the present one is that here we maximize average production rate instead of expected stochastic flexibility (Straub & Grossmann, 1990). The system resides during a portion of the operating horizon in states where some of the components are affected by failure modes that temporarily decrease production rate. Maximizing average production rate over a long operating time involves designing the system so that the effect of random failures is minimized and, therefore, has a similar effect on the system design as the objective of maximizing E(SF). Computation of E(SF) relies on the criterion of whether or not the system can match the demand rate in each of the discrete failure states and fails to capture the difference between a state where the production rate is slightly less than the demand rate from one where the entire system is shut down. The objective of maximizing long-term average production rate better matches industrial design criteria than the maximization of E(SF). Appendixof this paper contains the complete mathematical formulation. Details of the model and a description of each of the constraints can be found in Terrazas-Moreno et al. (2010).

STOCHASTIC PROGRAMMING REPRESENTATION

We model the problem of optimal design of an integrated site (IS) as a two-stage stochastic mixed-integer linear program. The vector of first stage design decisions d includes binary variables to represent the selection of production units from a superstructure and continuous variables, such as production unit capacities and storage tank sizes. Stage two decisions only involve continuous variables. A number of failure modes contained in a set L can occur in the production units of the superstructure at random times. This fact introduces endogenous uncertainty to the operation of the IS. Furthermore, this uncertainty is of a discrete nature (whether or not failure $l \in L$ occurs) and can be modeled using a parameter yl that is 0 if failure l occurs and 1 otherwise. The probability of a failure l occurring at any point in time is probl. We define a vector $y = \{yl\}$ where $l = 1, ..., |L|$; this vector has zeros in the positions corresponding to failures occurring simultaneously at any given time in the IS. There is a finite number of possible 0–1 combinations for the vector y. Each of these combinations defines a scenario in set S. Therefore, each scenario $s \in S$ corresponds to an instance of the vector y , and it can be represented as $y_s = \{y_s^{\ell}\}$, $l = 1, ..., |L|$. Assuming independent probabilities of failures, the probability associated with each scenario is $p_s = \prod_{\ell : y_s^{\ell} = 0} prob^{\ell} \prod_{\ell : y_s^{\ell} = 1} (1 - prob^{\ell})$. The second stage variables $x_s \in R q$ are used to model material flows in the integrated site (IS) for each scenario $s \in S$.

Remark

Set S contains all possible failure scenarios in the superstructure, but each flowsheet selection defines a subset of relevant scenarios. The scenarios in S can be aggregated – several scenarios can be projected into a single one – to derive any relevant subset for any possible flowsheet selection.

The following mixed-integer linear programming (MILP) problem is the deterministic equivalent of the stochastic optimal design problem. It also corresponds to a compact representation of the model presented in Appendix A. In the MILP model below, the variables x_s represents the vector of second-stage decision variables defined in the model in Appendix A. These variables are $flow_{j,j',n,s}, f^P_{j,n,s}, f^B_{j,n,s}, f^{IN}_{j,n,s}, f^{OUT}_{j,n,s}, ps_{m,s}, \delta^s_{j,n}$. The variable d is a vector of first-stage design variables that corresponds to the following variables in Appendix A: $inv_{j,n}, s^-d_{j,n}, v_{j,n} pc_m, z_m$.

$$Max \sum_{s \in S} p_s c_s^T x_s - pen^T sl \tag{1a}$$

s.t.

$$A_s^1 x_s + B_s^1 d \le c^1 \quad \forall s \in S \tag{1b}$$

$$A_s^2 x_s \le diag(B_s^2 d(e - y_s)^T) \quad \forall s \in S \tag{1c}$$

$$\sum_{s \in S} F_s x_s + Gd - sl \le h \tag{1d}$$

$$C_3 d \le Capital \tag{1e}$$

$$\begin{bmatrix} B(d)_{s,s'} \\ x_s = x_{s'} \end{bmatrix} \vee \begin{bmatrix} \neg B(d)_{s,s'} \\ x_s \ge 0 \\ x_{s'} \ge 0 \end{bmatrix} \quad \forall s \in S, s' \in S, s' < s, (s, s') \in NA \tag{D1}$$

$$d \in D, x^s \in \Re^q, sl \in \Re_+, \quad where \quad D = \{d | d_i = 0, 1,$$

$$i = 1, i = 1, \ldots, p, d_i \in \Re_+^r, i = p+1, \ldots, r\}$$

$$B(d)_{s,s'} = \{True, False\} \quad \forall (s, s') \in S^2 \tag{1h}$$

The objective function (1a) represents the maximization of the average flow of product to external consumers. The coefficient *penT* penalizes the slack variables *sl* that are introduced in constraint (1d) for guaranteeing feasibility of the subproblems and accelerating the convergence of our implementation of Benders decomposition. The penalty term is large enough to enforce the slack variable to become equal to zero at the optimal solution. In the examples we solve in this paper, we use a value of 10 for the penalty term. Constraints (1b) and (1c) represent the mass balances in the superstructure. Constraint (1c) imposes reductions on the production rates of the plants in the integrated site, according to the active failure modes in

state s ; it will have terms of the form: $ps_{m,s} \le pc_m [1 - (1 - y_\ell^s)(rc_\ell)]$ (refer to constraint (A11) in Appendix A). The notation *diag(X)* corresponds to the diagonal elements of the matrix *X*; and *e* is a unitary vector. Constraint (1d) is a compact representation of the intermediate storage model we proposed in previous work (Terrazas-Moreno et al., 2010). Constraint (1e) corresponds to the ε-constraint for the bi-criterion optimization problem since it restricts the cost of a flowsheet with design variables d to be less than or equal to the maximum available investment, *Capital*. Disjunction (D1) establishes a relationship between the operating variables of scenario s and s '. The Boolean variable $B_{(d)s,s'}$ is true if scenarios s and s' are indistinguishable in the flowsheet defined by d. It is important to notice that (D1) is not defined for every possible pair (s',s), but only for those pairs that fulfill two conditions: (a) by symmetry of the constraint set, the first condition is $s' < s$; (b) that pair (s',s) should belong to set NA, which is made up of all pairs that differ in the value of exactly one element of the vector ys. These properties are explained in detail by Goel and Grossmann (2006). We can represent (D1) using inequality constraints and a binary variable $\alpha_{s,s'}(d)$ in (1f) and (1g) (Raman & Grossmann, 1994), which yields the following MILP problem:

$$Max \sum_{s \in S} p_s c_s^T x_s - pen^T sl$$

s.t.

$$A_s^1 x_s + B_s^1 d \leq c^1 \quad \forall s \in S$$

$$A_s^2 x_s \leq diag(B_s^2 d(e - y_s)^T) \quad \forall s \in S$$

$$\sum_s F_s x_s + Gd - sl \leq h$$

(1d')

$$C_3 d \leq Capital$$

$$x_s \leq x_{s'} + M\alpha_{s,s'}(d) \quad \forall s \in S, \quad s' \in S, \quad s' < s, (s,s') \in NA$$ (1f)

$$x_s \geq x_{s'} - M\alpha_{s,s'}(d) \quad \forall s \in S, \quad s' \in S, \quad s' < s, (s,s') \in NA$$ 1g)

$d \in D, x^s \in \Re_+^q, sl \in \Re_+$ where $D = \{d | d_i = 0, 1 ,$

$\qquad i = 1, \ldots, p, d_i \in \Re^r, i = p+1, \ldots, r\}$

$\alpha_{s,s'}(d) = 0, 1 \quad \forall(s, s') \in S \times S$ \hfill (1h')

Inequalities (1f) and (1g) are non-anticipativity constraints, where the constant M is a large number that renders these inequalities to be redundant for $\alpha_{s,s'}(d)=1$. The term $\alpha_{s,s'}(d)$ is a function of the integrated site flowsheet and is defined as follows:

$$\alpha_{s,s'}(d) = \begin{cases} 0 & \text{if scenarios } s \text{ and } s' \text{ are indistinguishable in} \\ & \text{the network topology defined by } d \\ 1 & \text{otherwise} \end{cases}$$

For instance, if design d does not include unit m from the superstructure, then states s',s that differ only on whether or not unit m has failed are considered indistinguishable. The explicit function for $\alpha_{s,s'}(d)$ is as follows:

$$\alpha_{s,s'}(d) = \sum_{m \in M} \sum_{\ell \in L_m} z_m \sigma_\ell^{s,s'} \quad \forall j \in J, \ s \in S, \ s' \in S, \ s > s', (s, s') \in NA$$

\hfill (1i)

where $\sigma_\ell^{s,s'}$ is a problem *parameter* that can be derived from the vector y_ℓ^s. The vector y_ℓ^s is a fixed parameter defined in a previous section to be 1 if failure l does not occur as part of state s, and 0 otherwise. Let the parameter $\sigma_\ell^{s,s'}$ be defined as below.

$$\sigma_\ell^{s,s'} = \max\{y_\ell^s - y_\ell^{s'}, (1 - y_\ell^s) - (1 - y_\ell^{s'})\} \quad \forall \ell \in L,$$

$$s \in S, \ s' \in S, \ s > s'(s, s') \in NA$$

\hfill (1j)

In Eq. (1j) $\sigma_\ell^{s,s'}$ is set to one if states s and s' are distinguishable with respect to failure l; that is, if the failure occurs in one state and not in the other.

Remarks

- The state space S usually has high dimensionality, so that the number of constraints defined by $\forall s \in S, s' \in S, s' < s, (s, s') \in NA$ can become computationally intractable.

- For a fixed design $\hat{d}$, constraints (1f) and (1g) can be solved outside of the optimization problem (1). In this case, we can aggregate all indistinguishable scenarios *a priori*, and generate a *reduced* set *SB*.
- Problem (1) with fixed $\hat{d}$ and a reduced set *SB* of scenarios corresponds to a linear programming (LP) problem. Since this LP problem excludes constraints (1f) and (1g), it results in a decrease of orders of magnitude in the number of constraints when compared against the full two-stage MILP stochastic problem.

DECOMPOSITION ALGORITHM

Basic Idea

The algorithm we propose in this section results in a significant reduction of the number of non-anticipativity (NA) constraints that are initially considered in problem (1). This algorithm can be used (as we do in our numerical examples) with other existing modeling techniques to reduce the number of NA constraints as inGoel and Grossmann (2006). NA constraints are required only because the optimal design is a degree of freedom. The failure scenarios are built considering all units in the *superstructure*, but not necessarily all of these units will be part of the optimal network topology. When two or more failure scenarios are different only with respect to a failure in units that are not part of the network, a non-anticipativity constraint has to be activated to make those scenarios identical.

Fig. 3 and Fig. 4 illustrate this point for a three unit example.

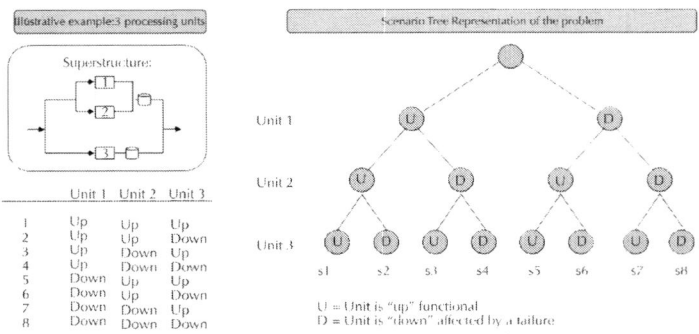

Figure 3: Scenario tree representation of failure scenarios.

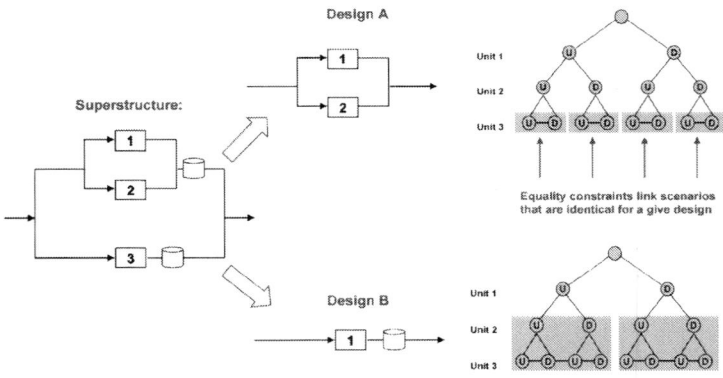

Figure 4: NA constraints are required since flowsheet structure is a degree of freedom.

The basic idea of the algorithm is to iteratively solve for the design (flowsheet) in a problem with only a few scenarios, and then solve the rest of the problem in a reduced space where only scenarios relevant to the fixed flowsheet are considered. Since only failures relevant to installed units are considered in the second step, there is no need for NA constraints. We use the basic concept of Benders decomposition to obtain the flowsheet in the master problem and to solve the rest of the scenarios in the subproblem. Our contribution to the method is that the subproblem is solved in a reduced space where there is a limited number of scenarios and no (or only very few) non-anticipativity constraints.

Definitions

According to what we have defined so far in the paper, we have the following sets:

$L=\{\ell:\ell$ is a failure mode in the superstructure$\}$

$S = \{s : s$ represents a combination of failure modes in $L\}$

$y^s = \{y^s_\ell : y^s_\ell = 0$ for active failure mode, 1 otherwise, $\ell = 1, \ldots, |L|\}$

We now include the following definitions

$\bar{L}_k \subseteq L$ is a subset of failures relevant to a network topology k

$S^k_B = \{s : s$ represent a combination of failure modes in $\bar{L}_k\}$

SM={s:subset of failure modes for master problem},   SM$\subseteq$S

$S^C_M :=$Complement of S_M

Finally, we define a subset of the Cartesian product as:

$$FN_k = \{(s', s) : y^{s'}_\ell = y^s_\ell, \quad \forall \ell \in \bar{L}_k\} \subseteq S \times S^k_B$$

The pair (s',s) represents all the duplets in the Cartesian product $\left(S \times S^k_B\right)$. For each of these duplets there is a corresponding pair ys',ys. Recall that ys is a vector with as many elements as failure modes l $\in$ L , and that $y^{s'}_\ell$ is a vector with as many elements as failure modes in l $\in$ Lk, where $|Lk| \leq |L|$ since there are more failures in the units of the superstructure that in the units corresponding to a particular topology k, which is conformed of a subset of the units in the superstructure. If for failures relevant to topology k, that is $l \in Lk\ y_\ell^{s'} = y_\ell^s$, scenarios s' and s are equivalent with respect to failures relevant to topology k. That is, looking only at the units in topology k, scenarios s' and s, have the same combination of active failure modes. If this condition is met, (s',s) are members of FNk.

In fact, FNk defines a function $FN_k : s \rightarrow S_B^k$ that maps between the sets S and S_B^k, since for every element $s' \in S$ there is only one corresponding element in $s \in S_B^k$. This relationship operates in the following way: taking any element in $s' \in S$ we look for the element $s \in S_B^k$ such that $(s', s) \in FNk$.

The function $FN_k : s \rightarrow S_B^k$ can be used to project the different scenarios in S that are identical with respect to the failures in Lk onto one scenario in S_B^k. This use of FNk is illustrated in Fig. 5.

$$FN_k = \left\{ (s',s) : y_\ell^{s'} = y_\ell^s, \forall \ell \in \overline{L}_k \right\} \subseteq S \times S_B^k$$

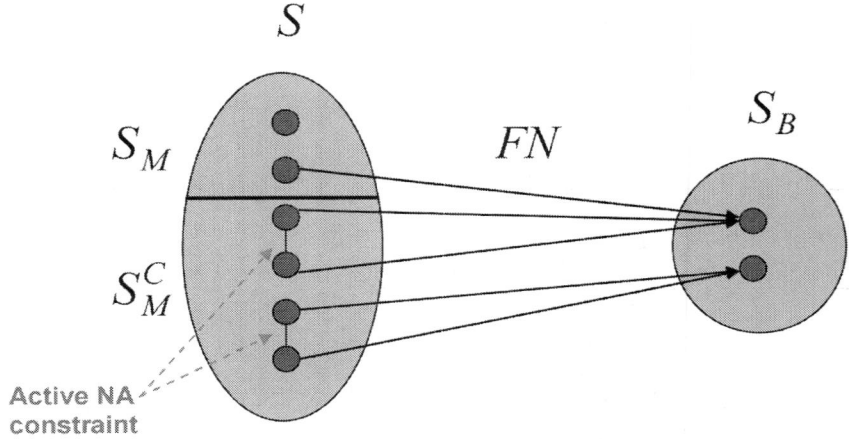

Figure 5: Function FNk.

PROPERTIES OF THE REDUCED SET S_B^k

In the above section, *SM* represents the failure modes considered in the master problem, while S_B^k was defined as a set of scenarios that include all the relevant failure mode combinations for a fixed network topology *k*. The main idea behind the decomposition approach presented in this paper is to solve the Benders subproblem in the reduced space of S_B^k. For instance, let the superstructure of an integrated site have two units. Each unit has one failure mode,

so $L = \{l_1, l_2\}$. Each vector $Y_s = \{y_s^{\ell_1}, y_s^{\ell_2}\}$ describes a scenario $s \in S$ where $S = \{\{1,1\},\{1,0\},\{0,1\},\{0,0\}\}$. Recall that a 0 in the vector *ys* denotes the occurrence of a failure in scenario *s*. Assuming the probability of failure is small, the Benders algorithm is set up

so that $SM = \{\{1,1\}\}$ and its complement $S_M^C = \{\{1,0\},\{0,1\},\{0,1\}\}$. If in a given iteration the flowsheet obtained from the solution of the master problem includes only one unit, we would define the

reduced set $\overline{L_k} = \{\ell_1\}$, so that $S_B^k = \{\{1\},\{0\}\}$. In this case scenario $\{1\}$ in S_B^k is the projection of $\{1,0\}$ from S onto S_B^k. Scenario $\{0\}$ is the projection of $\{0,1\},\{0,0\}$.

An important property that we require of S_B^k is that the sum of the probabilities of the scenarios in the reduced set must be equal to the summation of the probabilities in the original set, i.e.,

$\sum_{s \in S_B^k} p_s^k = \sum_{s \in S_M^k} ps$. We illustrate how to compute the probabilities $ps, \forall_s \in S_B^k$ in order to satisfy this condition using the example of a two-unit superstructure. There is one failure mode per unit in each of the two units, where $prob^{l_1} = 0.02$ and $prob^{l_2} = 0.03$. Fig. 6 shows the combinations of failure modes for scenarios in S and their corresponding probabilities. It also shows the partitioning of S into *SM* and S_M^k.

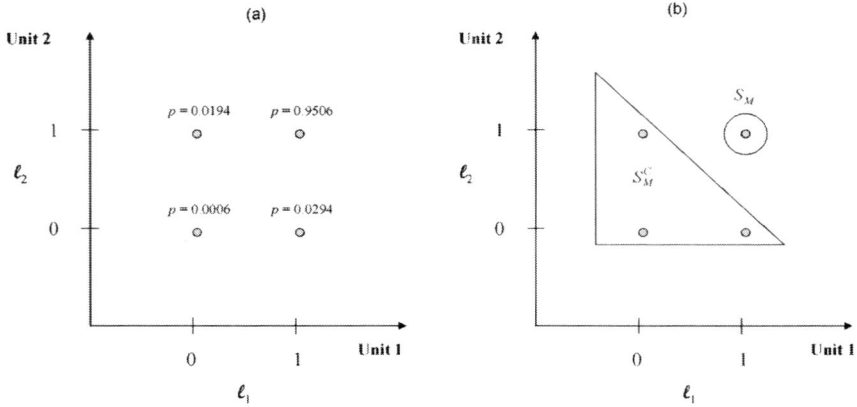

Figure 6: Scenarios and their partitioning in illustrative example.

Once more, assume that at a given iteration of the Benders algorithm only unit 1 is chosen from the superstructure, and we wish to solve the subproblem in the projected space S_B^k. There are two states in S_B^k: {1} and {0} corresponding to the functional and failed states of unit 1. The first of these, {1}, is the projection of {1,0}; the second, {0}, is the projection of {0,1},{0,0}. The probabilities of the reduced states {1} and {0} are equal to the sum of the probabilities of the projected states {1,0} and {0,1},{0,0}. The probability of the state {1} is equal to the probability of {1,0}, and the probability of {0} is the sum of the probabilities of {0,1},{0,0}. Fig. 7 shows the reduced scenarios and their corresponding probabilities for fixed unit 1. It can be verified that each of the probabilities in the scenarios in Fig. 7 correspond to the addition of probabilities of the first and second column of Fig. 6(a), considering only the states in S_M^C. We label the probability in the reduced space as $P_{s'}^k$. In general,

$$P_{s'}^k = \sum_{s \in A} ps_{,,} \text{ where } A = \left\{ s : y_s^\ell = y_{s'}^\ell, \forall l \in \bar{L}, s \in S_M^C \right\}.$$

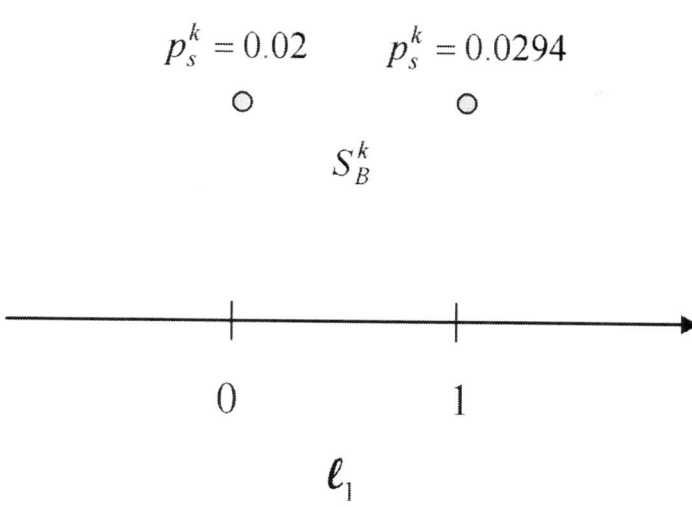

Figure 7: Reduced scenarios in S_B^k and their probabilities p_s^k for fixed unit 1.

Proposed Algorithm

Step 1

Define a maximum number of iterations K_{max} and the tolerance of the problem ε. Set the counter $K = 1$, and the initial value for the lower bound $LB = -\infty$.

Select *SM* as the set of scenarios with largest probability *ps* (i.e. $ps \geq \gamma$, where γ is threshold value).

Step 2

Solve the master problem (*M*) as defined below:

Max Constant $+ \varepsilon\eta$ (M1)

s.t.

$$\eta \leq \sum_{s \in S_M} p_s c_s^T x_s + \sum_{s \in \bar{S}_B^k}(u_{1s}^k(c^1 - B_s^{1^k}d) + u_{2s}^k diag(B_s^{2^k}d(e - y_s)^T))$$

$$+ v^k(h - Gd - \sum_{s \in S_M} F_s x_s) + \sum_{s \in S_B^{ks'}:s' \in S_M,} \sum_{(s,s') \in FN_k \cap NA} w_{s,s'}^k x_{s'},$$

$$K \neq 1, \quad k = 1, \ldots, K-1 \qquad\qquad (M2)$$

$\eta \geq LB + \varepsilon$ (M3)

$$\eta \leq \sum_{s \in S_M} p_s c_s^T x_s + \sum_{s \notin S_M} p_s c_s^T x_s^{UB}$$

(M4)

$$A_s^1 x_s + B_s^1 d \leq c^1 \quad \forall s \in S_M$$

(M5)

$$A_s^2 x_s \leq diag(B_s^2 d(e - y_s)^T) \quad \forall s \in S_M$$

(M6)

$$\sum_{s \in S_M} F_s x_s + Gd \leq h - \sum_{s \in S_M} F_s x_s^{LO}$$

(M7)

$C_3 d \leq Capital$ (M8)

$x_s \leq x_{s'} + M\alpha_{s,s'}(d) \quad \forall s \in S_M, \quad s' \in S_M, \quad s' < s, \quad (s,s') \in NA$ (M9)

$x_s \geq x_{s'} - M\alpha_{s,s'}(d) \quad \forall s \in S_M, \quad s' \in S_M, \quad s' < s, \quad (s,s') \in NA$ (M10)

$d \in D, x^s \in \Re_+^q, sl \in \Re_+ \quad where \quad D = \{d|d_i = 0, 1,$

$$i = 1, \ldots, p, d_i \in \Re^r, i = p+1, \ldots, r\}$$

$\alpha_{s,s'} = 0, 1 \quad \forall (s,s') \in S \times S$ (M11)

Where $u_{1s}^k, u_{2s}^k, v^k, and w_s^k$ are dual variables arising in the subproblems defined in Step 4?

The master problem (M) is the bottleneck of the decomposition algorithm. To speed up the convergence of (M), we have set it up as a feasibility problem instead of a rigorous optimization problem. The value of the constant term in the objective function (M1) is of the same order of magnitude as the optimal solution to the full space problem. This value is easy to calculate since it is possible to know the productivity if no failures were present in the integrated site. Then we solve (M) using a loose tolerance in the MILP solver. The tolerance that we refer to here is the gap between the upper and lower bounds *of the branch and bound method* used to solve the MILP. It is not to be confused with the tolerance defined for the Benders decomposition algorithm. For instance, if the tolerance of the Benders decomposition, which is ε in the nomenclature of this paper, is set to 2%, we can allow a gap of 5% for the branch and bound method. In (M1) the term *Constant* is much larger than the term $\varepsilon\eta$, so that (M) will converge once a feasible solution is found. Although, as we have just said "the term $\varepsilon\eta$ is comparatively small", keeping it is important since it improves the quality of the feasible solution found. Eqs. (M2) and (M3) constrain the variable η to be less than the dominant Benders cut but greater than a valid lower bound plus the convergence tolerance ε. Constraint (M4) has been added to enforce a valid upper bound on the objective function of the master problem. Note that the cardinalities of the sets of constraints (M9) and (M10) are much smaller than the cardinalities of (1f) and (1g). The solution to this problem yields the optimal values of the decision variables of the master problem at iteration k: $\hat{d}^k$ and $\hat{x}_s^k \forall s \in S_M$.

Termination criterion: Constraints (M5), (M6), (M7), (M8), (M9), (M10) and (M11) can always be trivially satisfied by not installing any unit from the superstructure and setting to zero the internal flows within the integrated site. The only possibility for (M) to be infeasible is if η cannot satisfy constraints (M2),(M3) and (M4). This is the case only if the upper bounds for η set by constraints (M2) and (M4) are lower than the lower bound set by constraint (M3). Therefore, *the algorithm is terminated as soon as (M) is infeasible*, which in turn guarantees that the lower bound *LB* is within ε-tolerance of the optimal solution to the full space problem.

Step 3

Select the sets $\bar{L}_k$ and S_B^k as defined above. Use the function *FNk* to compute the coefficients in the reduced space of S_B^k:

$$p_s^k = \sum_{s':s' \in S_M^C, \ (s',s) \in FN_k} p_{s'} \quad \forall s \in S_B^k$$

The calculation of this coefficient has been explained in the previous section.*Step 4*

Solve the subproblem (B) of the Benders decomposition using the reduced scenario set S_B^k

$$Max \sum_{s \in S_B^k} p_s^k c_s^T x_s^k + \sum_{s \in S_M} p_s c_s^T \hat{x}_s^k - pen^T sl \quad \text{(B1)}$$

s.t.

$$A_s^1 x_s \leq c^1 - B_s^1 \hat{d}^k \quad \forall s \in S_B^k \quad \text{(B2)}$$

$$A_s^2 x_s \leq diag(B_s^2 \hat{d}^k (e - y_s)^T) \quad \forall s \in S_B^k \quad \text{(B3)}$$

$$\sum_{s \in S_B^k} F_s x_s - sl \leq h - G\hat{d}^k - \sum_{s \in S_M} F_s \hat{x}_s^k \quad \text{(B4)}$$

$$x_s = \hat{x}_{s'}^k \quad \forall s \in S_B^k, \quad s' \in S_M, \quad (s,s') \in FN_k \cap NA \quad \text{(B5)}$$

$$x^s \in \mathfrak{R}_+^q, \quad sl \in \mathfrak{R}_q^+ \quad \text{(B6)}$$

The non-negative slack variables *sl* in (B4) ensure feasibility of the subproblem. A special note must be made regarding constraint (B5). The reduced set S_B^k is introduced in order to eliminate the need for non-anticipativity constraints (1f) and (1g) among the scenarios in S_M^C, but there are still indistinguishable pairs of scenarios (*s,s'*) where *s* belongs to S_M^C and *s'* to *SM*. The claim is that the set of constraints in (B5) is of significantly smaller cardinality than (1f) and (1g).

After solving subproblem (B), we construct the Benders cut using the objective function of the dual of (B):

$$\eta \le \sum_{s \in S_M} p_s c_s^T x_s + \sum_{s \in \bar{S}_B^k} (u_{1s}^k (e - B_s^{1^k} d) + u_{2s}^k diag(B_s^{2^k} dy_s^T))$$

$$+ v^k \left(h - Gd - \sum_{s \in S_M} F_s x_s \right) + \sum_{s \in S_B^k : s' \in S_M,} \sum_{(s,s') \in FN_k \cap NA} w_{s,s'}^k x_{s'}$$

Where $u_{1s}^k, u_{2s}^k, v^k, and w_s^k$ are the optimal dual multipliers of constraints (B2), (B3),(B4) and (B5) at iteration k.

The value of the lower bound (LB) is updated if the optimal value of (B) is greater than the incumbent LB.

If $K = K_{max}$, the algorithm stops. Otherwise, set $K = K + 1$, and go to *Step 2*.

DISCRETE RATE SIMULATION MODEL

The simulation begins with the generation of discrete items, each representing a unique failure mode. Each plant in the integrated site has one or more failure modes. For each failure mode, we look up which plant is impacted by the failure. We then look up the number of parallel units for that plant and the production rate per unit. These parameters are assigned to the failure mode as attributes so the look-up is necessary only once at the beginning. Each item is unbatched to N items, where N = number of parallel units. The underlying assumption is that if multiple parallel units exist in a plant, these units fail independently of each other. For each failure mode, its time between failure (TBF) and time to repair (TTR) are calculated from distributions developed from available data. All failure modes wait for failure start, for the duration corresponding to their respective TBF. Then they wait in a queue, and each is released only when the affected unit/train is running (no ongoing failure or turnaround). When a failure mode is released from this queue, the rate and status of the affected unit is updated. All failure modes wait for repair to finish, for the duration corresponding to their respective TTR or less (only if the repair is pre-empted by a simulation logic that synchronizes certain maintenance activities). After the repair is finished for a given failure mode, the rate and status of the affected unit/train are restored. The TBF and TTR are recalculated, and the process is repeated. Fig. 8 summarizes the simulation logic for modeling unplanned downtimes resulting from failures.

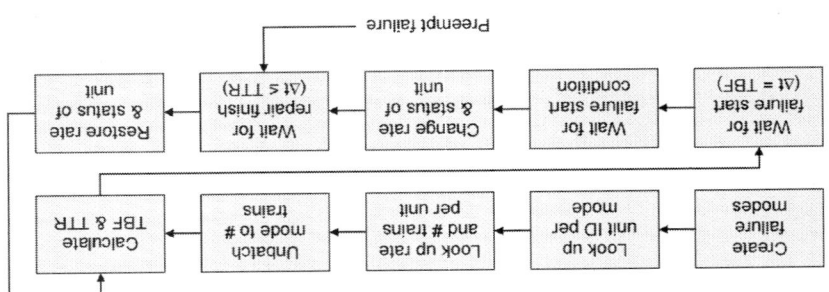

Figure 8: Simulation logic for unplanned downtimes.

A failure mode can result in the rate of zero for a complete shutdown or between zero and maximum unit capacity for rate loss. If there is an ongoing failure at a unit such that the current rate is zero, the above mentioned queue prevents any further failure modes for that unit to be activated, the assumption being that a down plant cannot fail any more. If there is an ongoing failure at a unit such that the current rate is not zero, then additional failure is possible. During the simulation, multiple failure modes can be in progress at the same time. The lowest rate of all failure modes in progress is chosen as the rate of a particular unit. The existence of more than one parallel unit per plant is accounted for in the updating of this rate input, depending on the status of all units. For example, if one unit is up and the other is down, then the rate input would be half the plant capacity.

APPLICATION TO THE DESIGN OF INTEGRATED SITES (IS)

Illustrative Example

We use the small example shown in Fig. 9 that was presented in our previous work (Terrazas-Moreno et al., 2010) to illustrate the proposed algorithm. The model and process data required to solve this example are available in Appendix A. We solve the full space version of the problem and then decompose it using the proposed algorithm for different values of capital investment in order to obtain a set of Pareto-optimal solutions. All results were obtained using the MILP solver CPLEX version 12.1, running on GAMS 23.3, with a 2.8 GHz Intel Pentium 4 processor and 2.5 GB RAM.

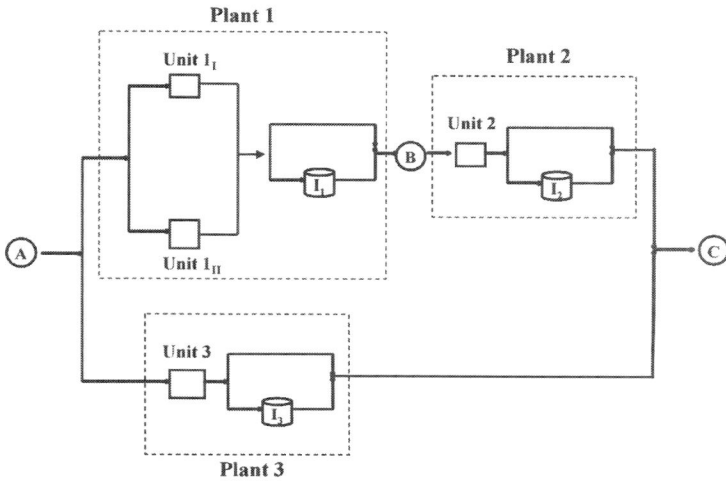

Figure 9: Integrated Site for the production of C from A.

The Pareto-optimal solutions are shown in Fig. 10. Two specific network structures are shown for the Pareto-optimal points A and B in Fig. 11.

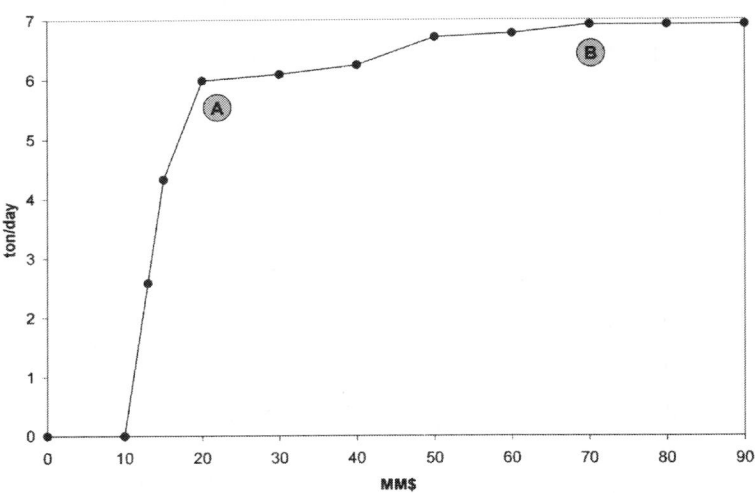

Figure 10: Pareto-set of optimal solutions for Example 1. (a) Configuration A. (b) Configuration B.

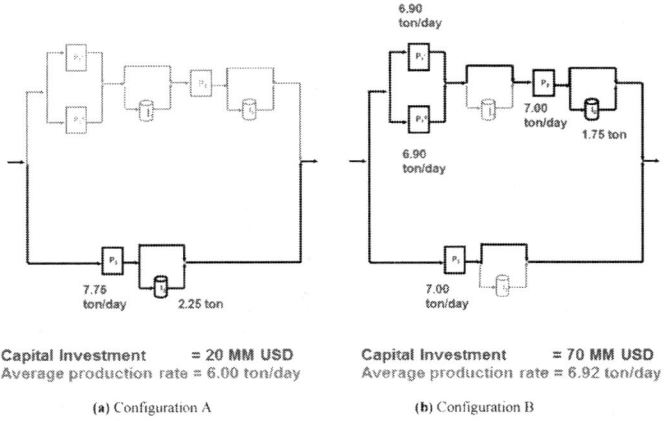

Figure 11: Network configurations for two Pareto-optimal solutions.

The network in Fig. 11(a) is designed to operate relying on process 3 and a large storage tank. In fact, 2.25 tons is the upper bound we set for the volume of the storage tank after plant 3. The network configuration that corresponds to point B in Fig. 10 belongs to a section of the optimal Pareto set where only marginal improvements in average production rate are achieved at the expense of large additional investment. Thus, the large spare capacities, the redundant units, and the relatively large storage tank inFig. 11(b).

Table 1 and Table 2 contain the problem sizes and the results that correspond to the full space model and to the decomposition strategy. Fig. 12 shows the convergence of the decomposition algorithm in 14 iterations for one value of capital investment (*Capital* = $60 MM). The lower bound before iteration 4 is a large negative number as a result of the slack variable in the objective function and Eq. (B4) having a large value. Without the slack variable the subproblem would have been infeasible. It is interesting to notice that, even for this small example made up of 4 processing units, the number of constraints in the full space model is significantly larger than the sum of the constraints in the master problem and subproblem (seeTable 1). This is a consequence of the reduction in the number of non-anticipativity constraints in (1f) and (1g), and in the number of scenarios in the subproblem. Table 2 shows the results obtained by the full space and Benders decomposition methods for different values of capacity investments. The example is so small that each Pareto-optimal solution can be obtained in a fraction of a second of CPU time either using the decomposition algorithm or directly solving the full space model.

Table 1: Statistics of illustrative example

	Discrete variables	Cont. variables	Constraints
Full space	4	1557	1670
Master problema	4	502	370
Subproblemb	0	265	177

[a]Master problem at iteration 1.

[b]Subproblem at iteration 1 for 13 MM USD of capital investment.

Table 2: Results of illustrative example

Capital investment (MM USD)	Optimal solution (ton/day)	
	Full space	Sub problem (2% tolerance)
13	2.61	2.59
15	4.35	4.33
20	6.10	5.99
30	6.10	6.09
40	6.25	6.25
50	6.76	6.71
60	6.77	6.77
70	6.92	6.92
80	6.92	6.92
90	6.92	6.92

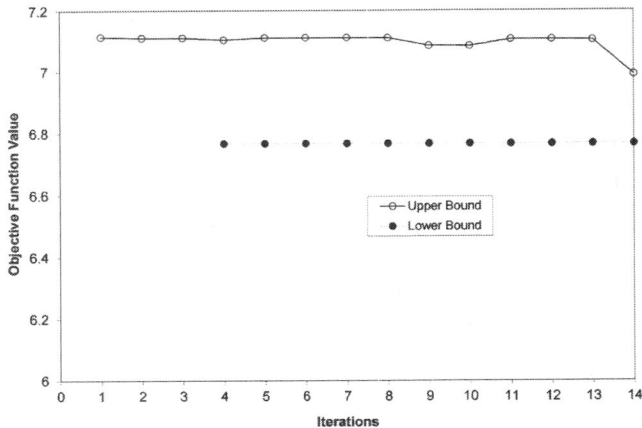

Figure 12: Iterations of master problem and sub problem in the illustrative example for $60 MM of investment.

Sensitivity Analysis

The Pareto-optimal solution that corresponds to point A in Fig. 10 represents an inflection point in the Pareto-optimal front. Due to the fact that further marginal increments in average production rate require large sums of extra capital investment, an industrial design team would be interested in understanding the factors that limit the performance of this design. For example, it could be more cost effective trying to improve the reliability characteristics of key components in the design, rather than adding parallel production trains or increasing the capacity of existing ones. It could also be the case that small changes in the volume of storage tanks or capacity of production units can yield significant improvements in average production rate. This information can be very useful with the optimization–simulation approach proposed in this paper. These teams could use it to search efficiently the design space around one or more Pareto-optimal solutions with discrete-rate simulation. The sensitivity analysis described below has the objective of providing this type of information to such a design team, so that it can be used as a guideline to fine-tune the design in a cost-effective way.

The first step to carry out the sensitivity analysis is to fix the selection of units in the superstructure to that in the Pareto-optimal point of interest. We then construct the set of scenarios relevant to this selection in the same manner as when setting up the subproblem in the decomposition algorithm. Using this collection of scenarios as set S, and *with a fixed selection of units* (fixed zm), we solve the full space problem described in Appendix A (Eqs. (A1)– (A22)) with the following two additional constraints.

$$pc_m^* \leq pc_m \leq pc_m^* \quad \forall m \in M \tag{S1}$$

$$v_{j,n}^* \leq v_{j,n} \leq v_{j,n}^* \quad \forall j \in J, \quad n \in N \tag{S2}$$

Where pc_m^* and $v_{j,n}^*$ are the capacity of unit m and the volume of the tank for product n after plant j in the Pareto-optimal design being analyzed. Any solver for LP problems, such as CPLEX, provides the reduced costs at the optimal solution. The reduced costs, which are the dual variables at the active bounds, correspond to the derivatives of the objective function with respect to perturbations on the right-hand sides of each of the constraints in the optimization formulation (Chvatal, 1983). By reading reduced costs of constraints (S1) and (S2), we obtain the values of the following derivatives:

$$\frac{\partial(\text{Average Production Rate})}{\partial(pc_m^*)} \quad \forall m \in M \tag{S3}$$

And

$$\frac{\partial(\text{Average Production Rate})}{\partial(v_{j,n}^*)} \quad \forall j \in J, \quad n \in N \tag{S4}$$

These derivatives show the sensitivity of the optimal solution for marginal increments in the design variables. In this way, the relative magnitudes of the sensitivities can be used as guidelines for fine-tuning the design using simulation tools.

Next, we are interested in finding the derivatives of the objective function with respect to the probabilities of each failure mode, that is,

$$\frac{\partial(\textbf{Average Production Rate})}{\partial(p_\ell)} \quad \forall \ell \in L \tag{S5}$$

This information can be used to determine key failure modes and look for ways to improve the reliability characteristics of the corresponding components in the design. The probabilities of being in an operational state with respect to failure l, p_l, do not appear explicitly in the MILP model defined by Eqs. (A1)–(A22). However, the probability of each failures scenario (discrete state), *probs*, which is a function of the probabilities of independent failure modes, does appear in constraints (A1), (A12) and (A13). Note that we assume that vrtsfrs=probs in constraint (A13), which is true when failures follow exponential distributions but an approximation in any other case. Knowing this, we carry out the following computations where *APR* stands for average production rate:

$$\frac{\partial(APR)}{\partial(p_\ell)} = \sum_{s \in S} \frac{\partial(APR)}{\partial(prob^s)} \frac{\partial(prob^s)}{\partial(p_\ell)},$$

(S6)

Where

$$\frac{\partial(APR)}{\partial(prob^s)} = \sum_{j \in J} \sum_{n \in N} \left[\frac{\partial(APR)}{\partial(A12_{j,n})} \frac{\partial(A12_{j,n})}{\partial(prob^s)} + \frac{\partial(APR)}{\partial(A13_{j,n})} \frac{\partial(A13_{j,n})}{\partial(prob^s)} \right]$$

$$+ \frac{\partial(A1)}{\partial(prob^s)}$$

(S7)

And

$$\frac{\partial(prob^s)}{\partial(p_\ell)} = \prod_{\ell':\{y^s_{\ell'}=1\},\ \ell' \neq \ell} p_{\ell'} \prod_{\ell':\{y^s_{\ell'}=0\}} (1 - p_{\ell'}) \quad \text{if } y^s_\ell = 1$$

(S8)

Or

$$\frac{\partial(prob^s)}{\partial(p_\ell)} = \prod_{\ell':\{y^s_{\ell'}=1\}} p_{\ell'} \prod_{\ell':\{y^s_{\ell'}=0\},\ \ell' \neq \ell} (1 - p_{\ell'}) \quad \text{if } y^s_\ell = 0$$

(S9)

In Eq. (S7), A12 and A13 stand for the right-hand sides of Eqs. (A12) and (A13) in Appendix A, and A1 corresponds to the objective function. The values $(\partial(APR)/(\partial(A12j,n))$ and $(\partial(APR)/(\partial(A13j,n))$ are the dual variables of the corresponding constraints, and they are obtained from the output of an LP solver. The partial derivatives $(\partial(A12j,n)/(\partial(probs))$, $(\partial(A13j,n)/(\partial(probs))$ and $(\partial(A1)/(\partial(probs))$ are obtained analytically from constraints (A12) and (A13) and the objective function (A1). Finally, Eqs. (S8) and (S9) are the result of differentiating the function $prob^s \prod_{\ell:\{y^s_\ell=1\}} p_\ell, \prod_{\ell:\{y^s_\ell=0\}} (1-p_\ell)$ that determines the probability of each failure state as a combination of the probabilities of the independent failure modes.

Table 3 contains the results of the sensitivity analysis of the design labeled as A in Fig. 10. Note: Design A only involves plant 3.

Table 3: Sensitivity analysis of design corresponding to $20 MM[a]

$\partial(APR) / \partial\left(v_{3,C}^{*}\right)$	~0
$\partial(APR) / \partial\left(pc_{3}^{*}\right)$	~0
$\partial(APR)/\partial(p_{3})$	6.98

This design only includes unit 3 from the superstructure.

The results in Table 3 indicate that marginal changes in the design variables around their optimal values have little effect on the objective function. In fact, this is consistent with the small slope after point A in Fig. 10. In contrast, increasing the probability of being in an operational state of plant 3 has a large potential impact on the average production rate. To give the value of the derivative $(\partial(APR)/(\partial(p_{3}))$ a more tangible meaning, we can calculate the effect of a 5% increase in the ratio of MTBF over MTTR for plant 3. After some algebraic manipulation of the expression $p_{3} = (MTBF_{3}/(MTBF_{3} + MTTR_{3}))$ we get:

$$\Delta APR \ (5\% \ increase \ in \ MTTF/MTTR)$$

$$= \left[\frac{1.05}{1 + 0.05p_{\ell}} - 1\right]\frac{\partial(APR)}{\partial(p_{3})} = 0.044$$

Thus, an increase of 5% on the mentioned availability could result in the design corresponding to $20 MM going from 6.095 mass/h to 6.139 mass/h. This option would be preferred over increasing the capital investment from $20 MM to over $30 MM.

LARGE-SCALE EXAMPLE

This section describes the computational results of the proposed algorithm for solving an industrial-sized process network. Fig. 13 shows the 9 processing plants that constitute this network. Each of the plants in this integrated site represents the production of a chemical that can be shipped to external markets or used as raw material in a downstream process. In the latter case, the integrated site could be part of a very large chemical production site made up of several "smaller" integrated sites. In this example we use a concave cost function (the common six-tenths rule (Biegler, Grossmann, & Westerberg, 1997) for capturing the effect of different processing capacities on the capital investment required by processing units. We use a piece-wise linear approximation in order to keep the linearity of the model.

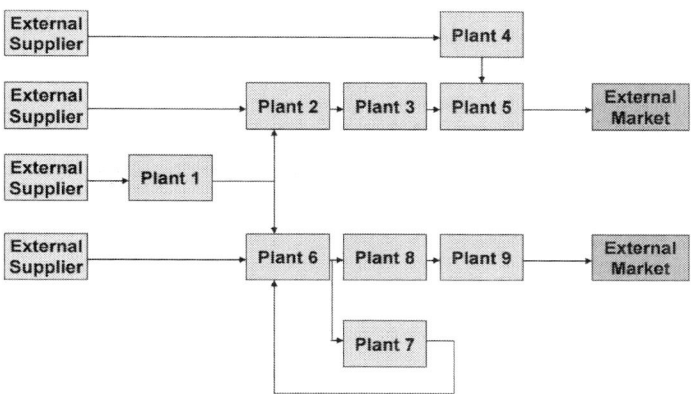

Figure 13: Industrial integrated site in large-scale example.

Each of the 9 plants that constitute the integrated site is modeled as a continuous process with one or several inputs and one output. We postulate a superstructure with two parallel production units in each plant and a storage tank after each plant except after plants 5 and 9, where the corresponding product cannot be stored. Each unit in the superstructure is subjected to different partial (decreased capacity) and total random failure modes. The distribution of the failure times is exponential, while the repair times follow a normal distribution. The total number of failures in the superstructure is 198.

Solving this industrial case study with the formulation in Appendix A using the decomposition framework presents two main challenges. One of them is the dimensionality: the resulting space for the units *in the superstructure* consists of 2^{198} discrete states (scenarios). Even with the decomposition algorithm, we are not able to handle this problem size. The second challenge is that the normal distribution of the repair times causes the variance of the residence time in each state to be different than that calculated assuming an exponential distribution. As indicated in our previous work (Terrazas-Moreno et al., 2010), the asymptotic values of the probabilities of the states and the mean residence time can be obtained analytically for any type of distribution as long as we have the mean failure and repair times. Unfortunately, this is not the case for the variance of the residence times. We overcome these two challenges by building a discrete event simulation model that we can use to collect a sample of scenarios. This technique is basically a Monte Carlo sampling procedure.

Sampling of Scenarios and Validation of Designs using the Discrete Event Simulation Model

One use of the simulation model is to build a sample of scenarios for our problem. As the simulation runs, it generates a list of items (failures) and their attributes (time between failures and time to repair). It also registers a generation time for each of the items (failures). Scenarios (states) in the problem at hand are described by a set of failures occurring simultaneously in the integrated site. From the list of items mentioned above, we can construct a list of scenarios or states and their durations. Appendix B contains the details of the methodology to construct the list of scenarios from the list of failure items created in discrete event simulation. Since the scenarios are generated randomly, we can consider the method as a Monte Carlo sampling. The number of scenarios is finite, and a few of them are highly more likely than the rest, so we can expect many repetitions of the same scenario as we sample. The sampling method allows us to calculate probabilities as well as mean and variance of residence time in each scenario. The size of the sample can be determined by the desired level of accuracy in the solution to our two-stage stochastic programming problem. The variance estimator of the solution to a stochastic programming problem is given by Shapiro and Homem-de-Mello (2000):

$$S(n) = \sqrt{\frac{\sum_{s=1}^{n}(E[obj] - obj_s)^2}{n - 1}}$$

(2)

where n is the number of scenarios and obj is the objective value. Let $z\alpha_{/2}$ be the normal standard deviate (corresponding to a normal distribution with zero-mean and unitary standard deviation) for a confidence interval of $1 - \alpha$. In other words, $Pr(z \le z\alpha_{/2}) = 1 - \alpha/2$, for $z \sim N(0, 1)$. For a confidence interval of 95%, $z\alpha_{/2} = 1.96$. To obtain the exact number of scenarios we follow the steps in You, Wassick, and Grossmann (2009):

- Obtain a sample for one year of operation to approximate $S(n)$.

- Define $z\alpha_{/2}$ and a desired interval H so that the solution to the stochastic programming problem is within an interval of [objsample $- H$, objsample $+ H$] with confidence of $1 - \alpha$

- Determine the size of the sample as:

$$N = \left[\frac{z_{\alpha/2}S(n)}{H}\right]^2$$

(3)

For $H = 2.5$ mass/h (recall that the objective function is the expected throughput of the integrated site) and a confidence level of 95% we need a sample size of 2878 scenarios. We found that by simulating 10 years of plant operation we can construct this sample. Intuitively, one would expect that excluding scenarios that are not relevant after 10 years of plant has a minimal effect on the optimal design of the integrated site.

Numerical Result

We simulated 10 year of operation of the integrated site and obtained a sample of 2973 scenarios. The statistics corresponding to the 10 most frequently encountered scenarios are shown in Table 4.

Table 4: Statistics of most probable failure scenarios

Scenario	Probability	Mean residence time (h)	Frequency of visits (1/(h × 10³))	Plants in failure mode
1	0.231	28	8.0	None
2	0.022	23	0.9	Plant 8
3	0.020	20	1.0	Plant 8
4	0.020	25	0.8	Plant 8
5	0.019	27	0.7	Plant 2
6	0.013	18	0.7	Plant 2
7	0.012	21	0.6	Plant 5
8	0.010	25	0.4	Pant 9
9	0.009	21	0.4	Plant 8
10	0.008	28	0.3	Plant 9

Table 4 contains some valuable information. For instance, from the first row we know that the integrated site will operate without any active failure mode around 23% of the time and that a failure will occur approximately every 28 h. Another important piece of information is that 4 out of the remaining 9 most common failure scenarios are due to some failure mode in plant 8.

After obtaining the sample of failure scenarios, we attempted to solve the problem in full space of 2973 scenarios (without decomposition) and using the proposed decomposition algorithm in a workstation with a 2.40 GHz Intel Core2 Quad CPU processor and 8 GB RAM. The results were obtained with the MILP solver CPLEX 12.1.0 running in GAMS 23.3.

The set of Pareto-optimal solutions in Fig. 14 was obtained first by solving the full space model which required 31 h. In contrast, using the decomposition algorithm with a 2% convergence tolerance, the Pareto curve can be obtained in 9 h. Table 5 shows the comparison in terms of CPU times for both approaches. An important note is that the solution times in the table do not include the time GAMS required to generate the model, which was nearly as long as the solution time for the full space model and not more than a few minutes with the decomposition algorithm. Table 6 shows the problem sizes of the decomposed and full space formulations. Finally, Table 7 shows the details of three of the Pareto optimal network configurations in Fig. 14.

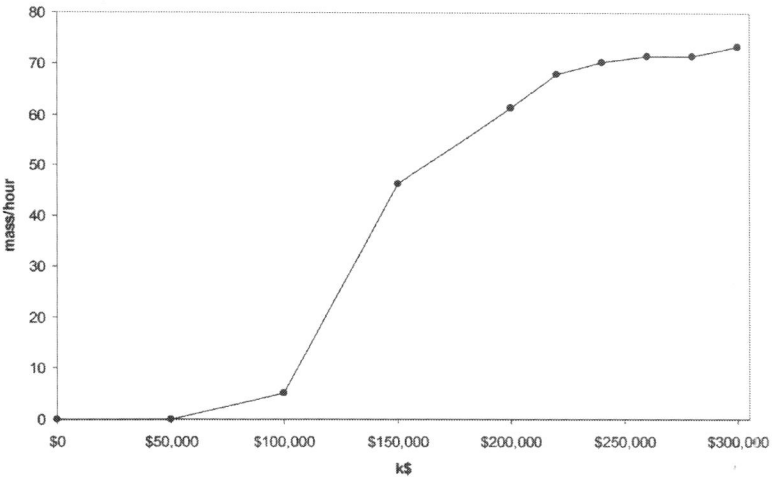

Figure 14: Pareto-optimal solutions for Example 2 using a random sample of 2973 scenarios.

Table 5: Performances of decomposition algorithm and full space solution using sample of scenarios

Capital investment (k USD)	Full space		Decomposition algorithm		
	Solution (mass/h)	CPU time (s)	Upper bound (mass/h)	Lower bound (mass/h)	CPU time (s)
$100,000	5.20	1,052	5.26	5.16	1,324
$150,000	46.83	4,980	47.28	46.35	6,431
$200,000	62.12	15,376	62.52	61.29	3,817

$220,000	68.30	23,271	69.27	67.91	3,813
$240,000	70.77	12,645	71.66	70.26	4,277
$260,000	72.31	21,611	72.97	71.54	3,308
$280,000	73.27	18,082	72.95	71.52	5,109
$300,000	73.97	15,245	74.92	73.46	3,080
	Total	31.18 h		Total	8.66 h

Table 6: Problem size corresponding to industrial case study

	Discrete variables	Cont. variables	Constraints
Full space	144	4,753,154	1,864,942
Master problem[a]	144	599,136	203,574
Subproblem	0	4,183,206	877,194

[a]Master problem at iteration 1.

Table 7: Optimal designs for different capital investments using a random sample of scenarios

Plant	200 MM USD Average Production Rate 61.29 mass/h			220 MM USD Average Production Rate 67.91 mass/h		
	Production units	Capacity per unit (ton/h)	Storage size (ton)	Production units	Capacity per unit (ton/h)	Storage size (ton)
1	1	74.8	–	1	88.2	–
2	1	77.1	–	1	87.8	–
3	1	58.7	–	1	64.4	3750
4	1	37.5	–	1	42.8	–
5	1	121.7	–	1	126.4	–
6	1	105.23	–	1	154.1	–
7	1	86.0	–	1	117.6	446
8	1	28.4	–	1	42.4	9
9	1	44.1	–	1	67.6	–

| Plant | 260 MM USD Average Production Rate 71.62 mass/h | | |
	Production units	Capacity per unit (ton/h)	Storage size (ton)
1	2	73.9	–
2	1	96.9	–
3	1	76.3	3750.0
4	1	42.8	3210.0
5	1	135.91	–
6	1	137.7	–
7	1	111.0	7770.0
8	1	42.4	93
9	2	52.0	–

Sensitivity Analysis

We carry out a sensitivity analysis around the design corresponding to $220 MM, following the same procedure outlined in Example 1. The analysis predicts that there is no marginal benefit in increasing the size of the storage tanks and that the design fine-tuning should focus on small increases on the capacity of plants 1 and 9. The sensitivity analysis with respect to probabilities in failure modes revealed that improving by 5% the ratio of MTBF over MTTR of one of the failure modes in either plant 2 or plant 8 could increase the average production rate from 68 ton/h to close to 70 ton/h. This is consistent with the data in Table 4where it can be seen that the most frequent failure modes occur in plant 8 and plant 2.

Simulation of the Pareto-optimal Designs

We use the same simulation model previously used for obtaining the sample of failure modes to evaluate the performance of the Pareto-optimal designs from the MILP optimization model. As we have indicated before, the simulation model is able to reproduce the real system to a greater extent than the optimization model, given the simplifications that have to be assumed to develop the MILP model.

Therefore, simulating the design obtained in the optimization step allows us to observe the average production rate (objective function of the optimization model) of the integrated site under more realistic assumptions than those incorporated in the constraints of the MILP model. Since we perform three simulations per configuration, we report average and standard deviation for each metric. Fig. 15 shows the Pareto-optimal curve obtained through optimization and the corresponding performance fixing the network configuration and capacities in simulation runs. The error bars above and below the points that correspond to simulation are set to one standard deviation obtained when running each point three times. As can be seen, the agreement between the optimization and simulation models is very good.

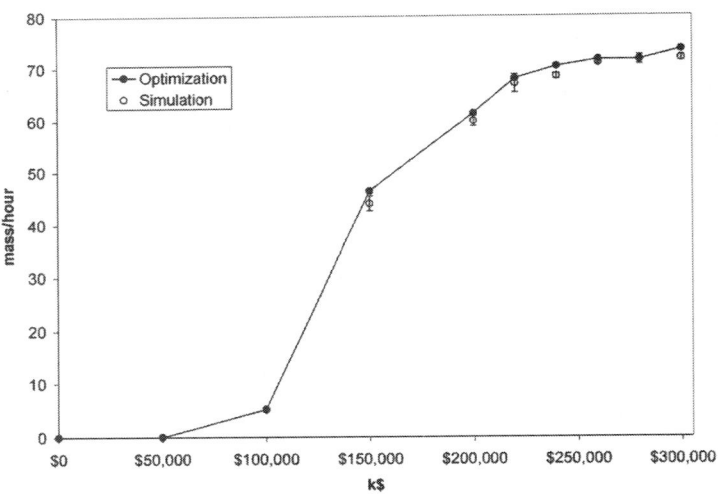

Figure 15: Comparison of simulation results vs. optimization results for some Pareto-optimal designs.

CONCLUSIONS

In this paper, we have addressed the problem of integrated site design under uncertainty as a two-stage MILP stochastic programming with endogenous uncertainties. In order to overcome the exponential growth in the number of scenarios required for modeling uncertainty, we proposed a decomposition algorithm based on Benders decomposition. Our main contribution is to exploit the problem structure in a way that allows the solution of Benders subproblems in a reduced space. The result is that the number of scenarios required is much smaller than with the full space problem. One of the main advantages of the proposed method is that the decomposed model requires significantly fewer non-anticipativity constraints.

The solution approach was tested in two case studies, one of them being an industrial processing network in which Monte Carlo Sampling was used to reduce the number of states. In the second case study, the decomposition algorithm reduces the solution time for the complete Pareto-set from 31 to 9 h. The integration with discrete rate simulation for validating/refining results increases the likelihood that the algorithmic method presented here will be accepted as a computational tool in an industrial setting.

ACKNOWLEDGMENTS

We acknowledge Jee H. Park from the Process Optimization group at The Dow Chemical Company for his contributions during project discussions, and The Dow Chemical Company for the financial support for this work. We also acknowledge the support of Imagine That, Inc. through the academic software grant.

REFERENCES

1.   Acevedo, J. & Pistikopoulos, E. N. (1998). Stochastic optimization based algorithms for process synthesis under uncertainty. Computers and Chemical Engineering, 22, 647–671.

2.   BASF. (2010). http://www.basf.com/group/corporate/en/ investor- relations/ basfbrief/verbund/ludwigshafen-site

Accessed on May 6th, 2010. Benders, J. F. (1962). Partitioning procedures for solving mixed variables programming problems. Numerische Mathematik, 4, 238–252.

3.   Biegler, L. T., Grossmann, I. E. & Westerberg, A. W. (1997). Systematic methods of chemical process design. New Jersey: Prentice Hall.

4.   Billinton, R. & Allan, R. N. (1992). Reliability evaluation of engineering systems. New York: Plenum Press.

5.   Chvatal, C. (1983). Linear programming. New York: W.H. Freeman and Company.

6.   Davies, K. M., & Swartz, C. L. E. (2008). MILP formulations for optimal steady-state buffer level and flexible maintenance scheduling. MASc Thesis. McMaster University. Dow. (2007). http://news.dow.com/dow news/corporate/2007/20070712a. htm Accessed on May 6th 2010.

7.   Ehrgott, M. (2005). Multicriteria optimization. Heidelberg: Springer Berlin. Geoffrion, A. M. (1972). Generalized Benders decomposition. JOTA, 10, 237–260.

8.   Goel, V. & Grossmann, I. E. (2006). A class of stochastic programs with decision dependent uncertainty. Mathematical Programming (Ser. B), 108, 355–394.

9.   Heyman, D. P. & Sobel, M. J. (1982). Stochastic models in operations research New York: McGraw-Hill.

10.   Ierapetritou, M. G., Acevedo, J. & Pistikopoulos, E. N. (1996). An optimization approach for process engineering problems under uncertainty. Computers and Chemical Engineering, 20, 703–709.

11.   ILOG. (2011). http://www-01.ibm.com/software/integration/ optimization/cplexoptimizer/ Accessed on January 15th 2011.

12.   Imagine That Inc. (2010). http://www.extendsim.com/index. html Accessed on March 23, 2011.

13.   Jonsbraten, T.W.,Wets, R.J. B. &Woodruff, D. L.(1998).Aclass of stochastic programs with decision dependent random elements. Annals of Operations Research, 82, 83–106.

14. Liu, C., Fan, Y. & Ordonez, F. (2009). A two-stage stochastic programming model for transportation network protection. Computers and Operations Research, 36(5), 1582–1590.

15. Miller, S., Owens, J. & Deans, D. (2006). From requirements development to verification – Designing reliability into a large scale chemical manufacturing system. In Reliability and maintainability symposium, 2006. RAM'06 annual (pp. 641–648).

16. Pistikopoulos, E. N. (1995). Uncertainty in process design and operations. Computers and Chemical Engineering, 19, S553–S563.

17. Pistikopoulos, E. N., Thomaidis, T. V., Melin, A. & Ierapetritou, M. G. (1996). Flexibility, reliability and maintenance considerations in batch plant design under uncertainty. Computers and Chemical Engineering, 20, S1209–S1214.

18. Pistikopoulos, E. N., Vassiliados, C. G., Arvela, J. & Papageorgiou, L. G. (2001). Interaction of maintenance and production planning for multipurpose process plants – A system effectiveness approach. Industrial and Engineering Chemistry Research, 40, 3195–3207.

19. Raman, R. & Grossmann, I. E. (1994). Modeling and computational techniques for logic based integer programming. Computers and Chemical Engineering, 18, 563–578.vadjust

20. Santoso, T., Ahmed, S., Goetschalckx, M. & Shapiro, A. (2005). A stochastic programming approach for supply chain network design under uncertainty. European Journal of Operational Research, 167, 96–115.

21. Shapiro, A. & Homem-de-Mello, T. (2000). On rate of convergence of Monte Carlo approximations of stochastic programs. SIAM Journal on Optimization, 11, 70–86.

22. Straub, D. A. & Grossmann, I. E. (1990). Integrated stochastic metric of flexibility for systems with discrete state and continuous parameter uncertainties. Computers and Chemical Engineering, 14(9), 967–985.

23. Straub, D. A. & Grossmann, I. E. (1993). Design optimization of stochastic flexibility. Computers and Chemical Engineering, 17(4), 339–354.

24. Terrazas-Moreno, S., Grossmann, I. E., Wassick, J. M. & Bury, S. J. (2010). Optimal design of reliable integrated chemical production sites. Computers and Chemical Engineering, 34(12), 1919–1936.

25. Van Slyke, R. & Wets, R. J. B. (1969). L-shaped linear programs with applications to optimal control and stochastic programming. SIAM Journal on Applied Mathematics, 17, 638–663.

26. Wassick, J. M. (2009). Enterprise-wide optimization in an integrated chemical complex. Computers and Chemical Engineering, 33(12), 1950–1963.

27. You, F., Wassick, J. M. & Grossmann, I. E. (2009). Risk Management for Global Supply Chain Planning under Uncertainty: Models and Algorithms. AIChE Journal, 55, 931–946.

An Optimization Model for Evaluating the Economic Impact of Availability and Maintenance Notions during the Synthesis and Design of A Power Plant

E. Godoy, S.J. Benz, and N.J. Scenna

CAIMI (Centro de Aplicaciones Informáticas y Modelado en Ingeniería), Facultad Regional Rosario, Universidad Tecnológica Nacional, Rosario, Argentina[1]

ABSTRACT

In this paper, we introduce an optimization strategy in order to comprehensively quantify the impact of availability and ma.ntenance notions during the early stages of synthesis and design of a new natural gas combined cycle power plant. A detailed state-space approach is thoroughly discussed, where influence of maintenance funds on each component's repair rate is directly assessed.

In this context, analysis of the reliability characteristics of the system is centered at two designer-adopted parameters, which largely influence the obtained results: the number of components which may fail independently at the same time, and the number of simultaneous failure/repair events.

Then, optimal solutions are evaluated as the availability-related parameters and the amount of resources assigned for maintenance actions are varied across a wide range of feasible values, which enable obtaining more accurate and detailed estimations of the expected economic performance for the project when compared with traditional economic evaluation approaches.

INTRODUCTION

The synthesis and design of a power plant are determinant stages of its life cycle, as they expose diverse degrees of freedom which can be manipulated in order to achieve significant improvements in the overall project economics. In this context, availability and maintenance notions play a key role even in these early phases as they directly impact on the ability of the plant to fulfill the desired generation goal.

Therefore, a comprehensive approach should be implemented to account for the consequences in the performance of the power plant, of achieving a desired availability level while assigning given resources for maintenance actions to accomplish such requirement. This task has proven to be a challenging one within an optimization context due to the large space of feasible solutions that needs to be analyzed, considering the wide array of design and operative decisions that could potentially improve the economic indicators of the power plant.

Availability and Maintenance in Process Design

In order to consider the effects of availability and maintenance in the plant economics, [Goel et al., 2002] and [Goel et al., 2003] indicated that revenues and operative costs must be affected by the system inherent availability, while an exponential relationship between investment and availability is used to compute the capital cost of each piece of equipment (if the inherent availability of each piece of equipment is considered explicitly as a continuous decision variable). Nevertheless, it could result quite challenging to obtain real-world data on how inherent availability is linked to capital cost for a given process equipment.

Frangopoulos and Dimopoulos (2004) introduced reliability aspects in the thermoeconomic model of a cogeneration system by means of the state-space method, so that redundancy is embedded in the optimal solution; thus obtaining more realistic values of the system profit. They applied such approach to the determination of the number of cogeneration packages necessary for achieving the desired production goal with a given availability level. In a three levels optimization problem, including synthesis and design, operation under time-varying conditions, and operation under partial failure, characteristics of a cogeneration plant were compared when solving the optimization formulation with and without reliability considerations. If reliability is taken into account, they observed that an extra cogeneration package is necessary in order to satisfy the minimum availability requirements, and proved that profits are overestimated when reliability aspects are ignored.

Luo et al. (2013) presented a methodology to minimize the total cost under normal conditions while reserving enough flexibility and safety for unexpected equipment failure conditions for the interconnected steam power plants that supply utility energy to a petrochemical complex. The proposed optimization strategy transforms the unexpected equipment failure scenarios into virtual periods which are inserted in between the normal scenarics, thus minimizing the total cost for real periods and reserving enough redundancy for the virtual periods (even though it requires a set of extra constraints for handling these last ones).

Aguilar et al. (2008) incorporated reliability and availability into the design (configuration and redundancy allocation) and operation (maintenance schedule) of flexible utility plants; and observed that two different tradeoffs may arise: capital investment versus contractual penalties for not fulfilling the power demand (which can be computed as profit losses during the plant down time); and capital investment versus costs originated by different failure scenarios (while evaluating if the plant is able to cope with the demand). It is observed that both tradeoffs present the advantage that can be implemented by using data commonly available in the literature and industry.

Haghifam and Manbachi (2011) concluded that improving repair rates exhibit a direct relation with the number of operators and their ability to undertake repairs, which is directly related with the annual budget assigned to maintenance. In their analysis, Iyer and Grossmann (1997) also computed the changeover costs for startup/shutdown of units between periods of operation. Moreover, these costs are the ones that link the different periods through binary variables. According to Tan and Kramer (1997), costs for operation at a derated state can be included in the analysis if a model can be determined for component degradation as a function of time, if quality can be modeled as a function of component performance, and if revenue can be modeled as a function of quality.

Erguina (2004) described a prototype model for nuclear power plant maintenance economics, aiming at understanding the impact of maintenance funds allocation on reliability and performance considering the plant life cycle. In the prototype model, it is considered that allocation of funds for preventive maintenance actions has an asymptotic effect on the system reliability, as beyond a given point, no significant performance improvement will be achieved even if additional resources are assigned to such effort. Tan and Kramer (1997) stated that there are four feedback mechanisms used to control and improve equipment reliability in a manufacturing plant: (1) corrective maintenance and failed equipment restoration,

(2) development of preventive maintenance strategies for improving plant safety and economics, (3) predictive maintenance followed by implementation of preventive maintenance actions, and (4) design and/or operative modifications for reliability improvement.

Mc Leod et al. (2007) addressed the optimization of nuclear power plant designs based on global risk reduction, focusing in two aspects of the problematic: design considering components quality and redundancy levels, and maintenance scheduling and human reliability. Resolution of the generated problem was pursued by means of an evolutionary algorithm (i.e. a combination of evolutionary strategies and genetic algorithms), which allowed obtaining a balanced design where the future maintenance and test schedules are established from a risk point of view for minimum total cost.

Sun and Liu (2014) proposed a stochastic model for determining the system configuration and operating scheduling of a steam and power plant, considering equipment selection and operation mode (normal, standby, or failure), as well as including compensation adjustments and penalties. Then, the obtained system design is able to cope with both normal operation and emergency scenarios, while a more accurate representation of the economic factors is obtained.

In a short-term combined economic environmental dispatch problem, Gjorgiev et al. (2013) introduced a measure of the availability of the generating units present in the system by means of a risk index which is a function of the generating units power level. As result, the authors observed that an increase of the availability of power generation was followed by a small increase of the fuel cost and the gaseous emissions, due to the opposed nature of the newly added objective function with respect to the two commonly considered ones.

Space of Possible Operative States

Applying the state-space method consists of three steps, as identified by El-Nashar (2008): first, identification of functional and failure modes of the system by making an inventory of all possible states; second, establishment of rules for transition between states and formulation of the transition rate matrix; and third, evaluation of expected values of the interest variables, by using the states probabilities as weights. For Markov modeling, with n component lifetimes divided into d time discretizations, steady-state solution of the Markov model requires storing $d \cdot 2^n$ elements of M for matrix calculations, if only binary states are considered for each component. Independence of events allows decoupling the component lifetimes, although this assumption may not be valid for multiple interacting components (Tan and Kramer, 1997).

As observed by Lisnianski et al. (2012), multi-state models are widely used in the field of power system reliability assessment, since using two-state models for large generating units usually yield pessimistic appraisals. Then, they presented a multi-state Markov model for a coal power unit, where the transition rates between the different generating capacity levels of the unit are estimated based on field observations, by means of an embedded Markov chain. As consequence, they observed that the values of the associated reliability indices are different from those calculated for a long-term range, while events as scheduled outages or planned maintenance cannot be taken into account. A disadvantage of this method is that the designer should perform capacity quantization if the actual derated operating states of the power plant are unknown.

Terrazas-Moreno et al. (2010) proved that it is possible to optimize over the state-space in the last level of the tree, even if the optimal flowsheet contains less units than those present in the last level; as the states in the last level of the tree that are derived from the same node are identical in terms of feasibility/infeasibility, and also, the sum of the probabilities of the states in the last level of the tree is equal to the probability of the state from which they are derived. They also observed that there are two ways for reducing the number of possible states: first, the failures with the same related rate reduction that occur in the same plant can be aggregated as one equivalent failure; and in second place, only the most probable states can be considered (as for example, those ones that cover 99% of the long term operative horizon).

Vassiliadis and Pistikopoulos (2001) presented a *MINLP* optimization framework for deriving optimal maintenance policies in continuous process operations in the presence of parametric uncertainty. To overcome the highly non-linear and combinatorial nature of the resultant model, availability threshold values of components (which are the minimum acceptable values of their availability, and therefore, determine the time at which maintenance actions need to be performed) were used to propose a two-steps resolution strategy. In addition, to describe the possible states of the process, they used a (operative-failed) state-space, where each possible state for the process has a given probability of occurrence, related to the time-varying availability of each component (as determined by its inherent characteristics and the maintenance policies).

Failure of a component in a cooling, heating and power (BHCP) system may result in failure of a sub-system or the whole system, as observed by Wang et al. (2013). For such a large system, the authors proposed to divide it into three parts in order to apply the state-space method: electricity (either from the gas turbine or from the grid), heat (which can be recovered through the steam generator of produced at the auxiliary boiler), and cool (obtained by means of the electric and/or absorption chillers). This procedure allows to more easily obtain expressions for each sub-system failure and repair rates.

Aim and Outline

In this work, implications of considering availability and maintenance tradeoffs during the synthesis and design stages of a new project for a *NGCC* power plant are thoroughly analyzed by means of a flexible equations-oriented mathematical formulation which accounts for the most important economic characteristics of the system. Meanwhile, every solution here presented is an optimal one, obtained when successfully achieving the resolution of a *MINLP* mathematical optimization model.

A state-space approach is used to account for the availability of the generation plant, thus facilitating a complete overview of the system operative condition across the entire time horizon (here adopted as annualized one given by the standard operative time), which also enables computing the optimal economic performance of the project more accurately. Moreover, impact of the amount of resources assigned for maintenance actions is evaluated through its influence on the repair rate of each major component which constitutes the power plant. The resultant *MINLP* model efficiently provides economic optimal solutions considering every feasible scenario that the plant has to deal with.

In addition, benefits of this strategy are evaluated by comparison against a traditionally designed power plant (i.e. a generation facility which does not contemplate these notions, obtained according to [Godoy et al., 2010] and [Godoy et al., 2011]). In this context, it is here found that the proposed approach allows obtaining more realistic design options when farther exploring the space of optimal solutions.

FORMULATION OF THE ECONOMIC OPTIMIZATION PROBLEM: STATE-SPACE MODELING OF AVAILABILITY AND COMPONENT-BASED ASSESSMENT OF MAINTENANCE RESOURCES IMPACT

Process Configuration

The flow diagram for the *NGCC* power plant is presented in Fig. 1. A 2 *GTs* + 1 *ST* multi-shaft power plant is selected as the generation driver (note that the second gas turbine and its associated steam generator are not presented in this figure).

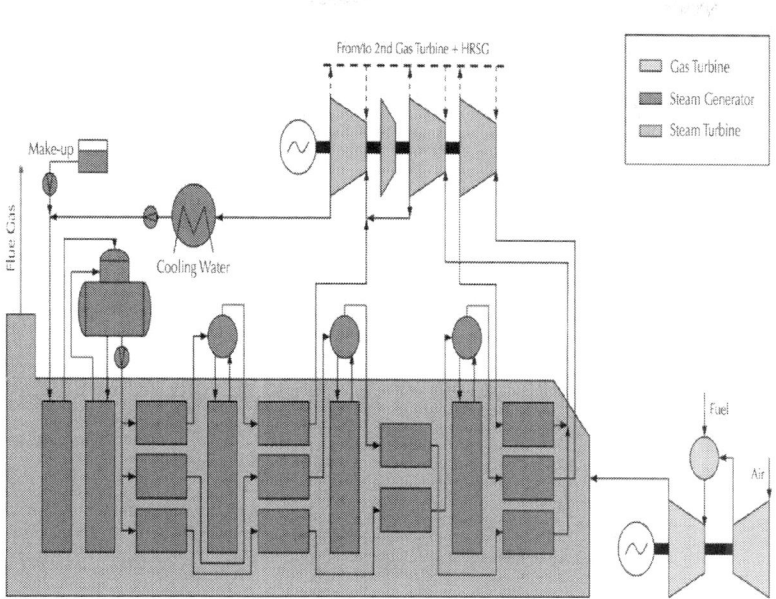

Figure 1: Flow diagram for the NGCC power plant.

Optimization Model

The mathematical formulation for the economic optimization of the NGCC power plant is presented in Fig. 2. In this optimization problem, the cost of the generated electricity is selected as objective

function $f(\underline{x}, \underline{y})$; which implies simultaneously minimizing the total expenditures of the project and maximizing the net energy output of the plant. Here, $\underline{x}$ are the sets of design and operative variables

and $\underline{y}$ are the sets of integer variables; while $\underline{h}(\underline{x}, \underline{y})$ and $\underline{\gamma}(\underline{x}, \underline{y})$ refer to the equality and inequality constraints which configure the mathematical model of the whole project.

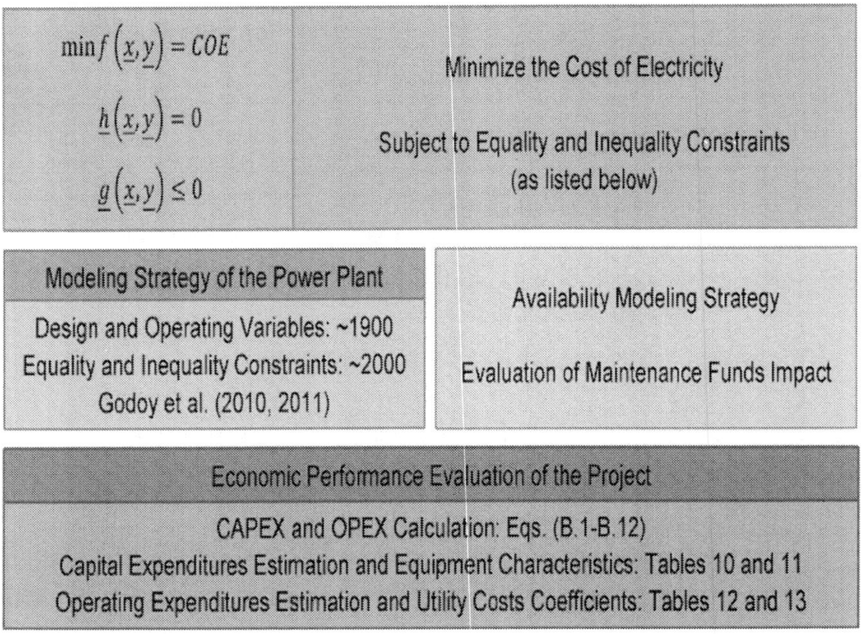

$$\min f\left(\underline{x},\underline{y}\right) = COE$$

$$\underline{h}\left(\underline{x},\underline{y}\right) = 0$$

$$\underline{g}\left(\underline{x},\underline{y}\right) \leq 0$$

Minimize the Cost of Electricity

Subject to Equality and Inequality Constraints
(as listed below)

Modeling Strategy of the Power Plant

Design and Operating Variables: ~1900
Equality and Inequality Constraints: ~2000
Godoy et al. (2010, 2011)

Availability Modeling Strategy

Evaluation of Maintenance Funds Impact

Economic Performance Evaluation of the Project

CAPEX and OPEX Calculation: Eqs. (B.1-B.12)
Capital Expenditures Estimation and Equipment Characteristics: Tables 10 and 11
Operating Expenditures Estimation and Utility Costs Coefficients: Tables 12 and 13

Figure 2: Economic optimization problem.

Objective Function: Cost of Electricity

The cost of the generated electricity *COE* gets computed according to Eq. (1) as the annualized cost *TAC* per unit of generated energy *GE*.

$$COE = \frac{TAC}{GE} = \frac{(CAPEX/CRF) + OPEX}{GE}$$

$$(1)$$

The annualized economic performance of the project is evaluated through its total cost, which includes capital expenditures *CAPEX* annualized by a given recovery rate *CRF*, and annual operative expenditures *OPEX*.

State-Space Modeling Framework

Number of Functional Modes and Their Interrelations

A functional mode is defined as the overall operative status of a system, which is here assumed as fully operational or failed. Each functional mode depends upon the set of operative statuses of its components, which impact on the ability of the system to fulfill its design purpose.

The number of functional modes and their interrelations depend upon the values of two parameters: the number of simultaneously and independently failed components N_{SIFC} and the number of simultaneous events N_{SE}. In order to illustrate this concept, Fig. 3 depicts an application example for a system constituted by three components, where it is observed that:

- Each possible functional mode is defined by the set of binary variables associated to every component, where an operative status is represented by a 1, while a 0 depicts a failed status

- The maximum number of zeros at each feasible functional mode is given by N_{SIFC}. Therefore, modes with 0 or 1 failed components may exist at state-space of Fig. 3(a) and (b), where a value of 1 is adopted for N_{SIFC}; meanwhile, modes with 0, 1 or 2 failed components may exist at state-space of Fig. 3(c) and (d), where a value of 2 is adopted for N_{SIFC}. Although not here graphically presented, a state-space with $N_{SIFC} = 3$ will additionally include a mode where all three components are failed (totalizing 8 feasible functional modes)

- Transitions between modes occur as the operative condition of one or more components change from operative to failed, or vice versa; where the maximum number of allowable simultaneous events is given by the value fixed by the designer for N_{SE}. When a value of 1 is adopted, transitions where only one component fails or is repaired are allowed, as shown in Fig. 3(a) and (c). If N_{SE} equals 2, the state-space will also include transitions between feasible modes which imply simultaneously changing the operative status of two components (i.e. two failures, two repair actions, or a failure and a repair action), as illustrated in Fig. 3(b) and (d). Similar conclusions are drawn for $N_{SE} = 3$

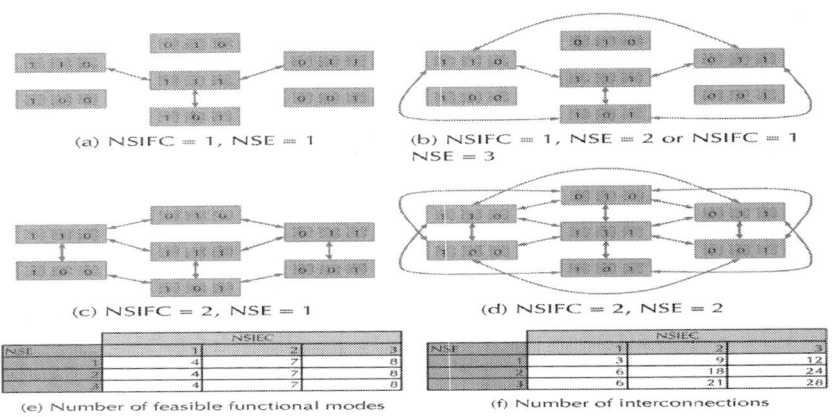

(a) NSIFC = 1, NSE = 1

(b) NSIFC = 1, NSE = 2 or NSIFC = 1
NSE = 3

(c) NSIFC = 2, NSE = 1

(d) NSIFC = 2, NSE = 2

(e) Number of feasible functional modes

(f) Number of interconnections

Figure 3: *NSIFC and NSE application example.*

For different values of N_{SIFC} and N_{SE}, Fig. 3(e) and (f) presents the number of feasible functional modes and the number of interconnections between them, respectively. It is here observed that:

- the number of feasible functional modes depends on the value adopted for N_{SIFC}, while is independent of N_{SE}
- the number of interconnections between modes increases as N_{SIFC} and N_{SE} do
- while N_{SIFC} is lower than the number of components, the state-space method is not able to represent every possible functional mode

Considering the aforementioned remarks, mathematical expressions for obtaining the space of feasible operative modes for the NGCC power plant, as function of N_{SIFC} and N_{SE}, are derived below.

Identification of Possible Functional Modes

The state-space method (i.e. a Markov-type approach, see for example Ibe (2009) and Kuo and Zuo (2003)) is used to evaluate the probability of the system being at each functional mode. These probabilities can be used for computing the operability indices of the system, as well as evaluating its technical and economical performance indicators.

The logically arranged subsystems, also referred as components, are listed in Eq. (2).

$$AS=\{AS_1,AS_2,...,AS_{NAS}\} \tag{2}$$

The number of components is determined by means of Eq. (3).

$$NAS=card(AS) \tag{3}$$

If only operative and failed statuses are considered for each of the components, the number of possible functional modes is given in Eq. (4).

$$NFM=2^{NAS} \tag{4}$$

Therefore, the space of the possible functional modes is listed in Eq. (5).

$$FM=\{FM_1,FM_2,...,FM_{NFM}\} \tag{5}$$

Then, a binary variable gets associated to each component, according to Eq. (6), in order to describe its status at each possible functional mode.

$$y_{AS,FM} = \begin{cases} 0 & failed \\ 1 & operative \end{cases} \tag{6}$$

The number of operative components at a given possible functional mode can be computed as the summation of the values of the binary variables for such functional mode, as described in Eq. (7).

$$N_{Op,FM} = \sum_{AS} y_{AS,FM} \tag{7}$$

Differentiation of Feasible Functional Modes

Functional modes can be differentiated according to a pre-specified criterion adopted by the designer, regarding the number of units that can independently fail at the same time $NSIFC$. Consequently, for a system with NAS components, the number of feasible functional modes is given in Eq. (8).

$$N_{FM^f} = 2^{N_{AS}} - \sum_{k=0}^{(N_{AS}-N_{SIFC})} \frac{N_{AS}!}{k!(N_{AS}-k)!}$$

(8)

The space of the feasible functional modes is listed in Eq. (9).

$$FM^f = \{FM_1^f, FM_2^f, \ldots, FM_{N_{FM^f}}^f\}, \quad FM^f \subseteq FM$$

(9)

The binary variable associated to each component, which describes its status at each feasible functional mode, is given in Eq. (10).

$$y_{AS,FM^f} = y_{AS,FM} \quad \textit{if} \quad N_{AS} - N_{SIFC} \leq N_{Op,FM} \leq N_{AS}$$

(10)

The number of operative components at a given feasible functional mode is given in Eq. (11).

$$N_{Op,FM^f} = N_{Op,FM} \quad \textit{if} \quad N_{AS} - N_{SIFC} \leq N_{Op,FM} \leq N_{AS}$$

(11)

Establishment of Transition Rules

Each component has a transition rate between its two statuses (operative and failed), as given by its failure and repair rates. Then, the component transition rate is given in Eq. (12).

$$z_{AS,FM_i^f,FM_j^f} = \begin{cases} \mu_{AS} & if \quad y_{AS,FM_i^f} = 0 \quad and \quad y_{AS,FM_j^f} = 1 \\ \lambda_{AS} & if \quad y_{AS,FM_i^f} = 1 \quad and \quad y_{AS,FM_j^f} = 0 \\ 0 & if \quad y_{AS,FM_i^f} = 0 \quad and \quad y_{AS,FM_j^f} = 0 \\ 0 & if \quad y_{AS,FM_i^f} = 1 \quad and \quad y_{AS,FM_j^f} = 1 \end{cases}$$

$$(12)$$

Failure and repair rates to be used in Eq. (12) are those associated to components defined in Eq. (2). Note that each component is constituted by several pieces of process equipment; so, their logical arrangement (series, parallel, redundancies) should be used to compute overall failure and repair rates associated to a given component ([NERC, 2011], [Participants, 2002] and [Alber et al., 1995]).

The overall transition rate from state FM_i^f to state FM_j^f is given by the transition rate matrix, as described in Eqs. (13) and (14), considering the number of simultaneous events NSE.

$$TRM_{FM_i^f,FM_j^f} = \begin{cases} \sum_{AS} z_{AS,FM_i^f,FM_j^f} & \forall i \neq j \quad if \quad 0 < \sum_{AS} |y_{AS,FM_i^f} - y_{AS,FM_j^f}| \leq N_{SE} \\ -TRMsum_{FM_i^f} & \forall i = j \end{cases}$$

$$(13)$$

$$TRMsum_{FM_i^f} = \sum_{FM_j^f} TRM_{FM_i^f,FM_j^f} \quad \forall i \neq j \quad if$$

$$0 < \sum_{AS} |y_{AS,FM_i^f} - y_{AS,FM_j^f}| \leq N_{SE}$$

$$(14)$$

Evaluation of Probabilities

The probability of the system being at every given feasible functional mode is obtained by solving the homogenous linear system of equations given in Eq. (15). Solving this system of equations implies finding the steady-state probabilities of an irreducible Markov process.

$$\sum_{FM_i^f} Pr_{FM_i^f} TRM_{FM_i^f,FM_j^f} = 0 \quad \forall j$$

(15)

An additional constraint is implemented when considering that the sum of state probabilities is always equal to one, as given in Eq. (16).

$$\sum_{FM_i^f} Pr_{FM_i^f} = 1$$

(16)

Expected Values of Variables and Operative Span

The expected value of a given variable can be computed as the weighted sum of the values for every feasible functional mode, as given in Eq. (17).

$$\hat{x} = \sum_{FM^f} Pr_{FM^f} x_{FM^f}$$

(17)

Within a given time horizon, the operative time associated to each functional mode is computed according to Eq. (18) as the standard operative time affected by the probability of occurrence of such functional mode. This follows from the fact that Pr_{FMf} can be interpreted as the average long-run proportion of the time that the system spends in the functional mode *FMf*.

$$POT_{FMf} = POT_0 Pr_{FMf}$$

(18)

Then, it is here observed that the state-space method accounts for every operative condition of the system across the entire time horizon (as probabilities of functional modes add up to 1). Note that the standard operative time POT_0 is here selected as the annualized time horizon, in order to account for the annualized plant maintenance schedule. Also, in this context, availability may be computed as the probability of the system of being at certain desired functional modes, where the process is able to fulfill the expected demand.

Description of Functional Modes of the *NGCC* Power Plant

The reliability block diagram associated to the power plant is introduced in Fig. 4.

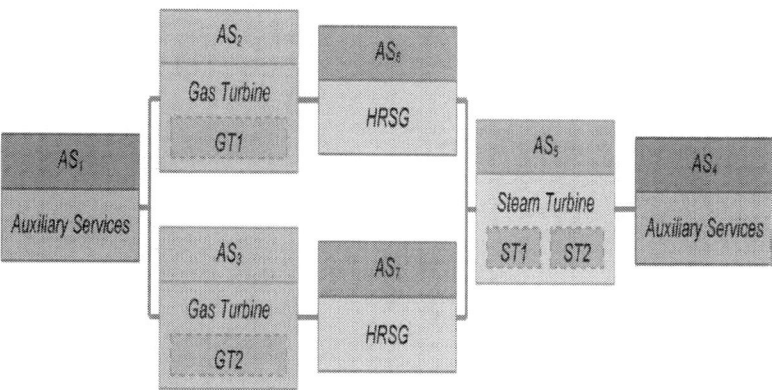

Figure 4: Reliability block diagram for the *NGCC* power plant.

At the power plant, the following components are identified according to Eq. (2) for purposes of availability analysis:

- Auxiliary services for both gas turbines (AS_1)
- Each gas turbine plus its associated generator (AS_2 and AS_3)
- Auxiliary services for both heat recovery steam generators and the steam turbine (AS_4)
- The steam turbine and its associated generator (AS_5)
- Each heat recovery steam generator (AS_6 and AS_7)

The number of availability-related components equals 7, according to Eq. (3). Then, the number of possible functional modes equals $2^7 = 128$, as stated in Eq. (4).

Overall Energy Generation and Resources Consumption for Each Functional Status

Altogether, the overall functional status of the power plant can be determined as the conjunction of the operative condition of every component, as defined in Eq. (19) and described in Table 2 regarding the delivered energy (and eventually, the consumed resources).

Table 2: Description of functional statuses for the *NGCC* power plant

Item	Description
P1	The power plant operates at full capacity
P2	Both gas turbines operate at full capacity, the steam turbine operates at half capacity
P3	Both gas turbines operate at open loop
P4	Only one gas turbine operates at full capacity, the other gas turbine is down, the steam turbine operates at half capacity
P5	Only one gas turbine operates at open loop, the other gas turbine and the steam turbine are down
P6	The power plant is down

$$FS=\{P1,P2,P3,P4,P5,P6\} \qquad (19)$$

Therefore, functional statuses of the power plant can be determined by specifying which of the sections, defined in Eq. (20) and introduced in Fig. 4, operate at full capacity, at a derated condition, or are down.

$$PP=\{GT1,GT2,ST1,ST2\} \qquad (20)$$

It is then noted that:

- The functional status of each gas turbine can be determined on its own, thus requiring one element associated to each of them (*GT1* and *GT2*)

- The description of the functional status of the steam turbine requires two elements (*ST1* and *ST2*), as it depends upon the steam generated at each *HRSG* and the operative condition of each associated gas turbine

A binary variable gets associated to each section, which describes its status at each feasible functional mode, as given in Eq. (21).

$$y_{PP,FM^f} = \begin{cases} 0 & failed \\ 1 & operative \end{cases}$$

(21)

As consequence, it becomes necessary to determine the functional statuses of the gas and steam turbines in terms of the operative condition of each component for every feasible functional mode, and afterwards transformed into the corresponding linear constraints, as described in Eqs. (22), (23), (24), (25), (26), (27), (28) and (29).

$$y_{GT1,FM^f} \geq y_{AS1,FM^f} + y_{AS2,FM^f} - 1$$

(22)

$$y_{GT2,FM^f} \geq y_{AS1,FM^f} + y_{AS3,FM^f} - 1$$

(23)

$$y_{GT1,FM^f} \leq y_{j,FM^f}, \quad j = AS1, AS2$$

(24)

$$y_{GT2,FM^f} \leq y_{j,FM^f}, \quad j = AS1, AS3$$

(25)

$$y_{ST1,FM^f} \geq y_{AS4,FM^f} + y_{AS5,FM^f} + y_{AS6,FM^f} + y_{GT1,FM^f} - 3$$

(25)

$$y_{ST1,FM^f} \geq y_{AS4,FM^f} + y_{AS5,FM^f} + y_{AS6,FM^f} + y_{GT1,FM^f} - 3$$

(26)

$$y_{ST2,FM^f} \geq y_{AS4,FM^f} + y_{AS5,FM^f} + y_{AS7,FM^f} + y_{GT2,FM^f} - 3$$

(27)

$$y_{ST1,FM^f} \leq y_{j,FM^f}, \quad j = AS4, AS5, AS6, GT1$$

(28)

$$y_{ST2,FM^f} \leq y_{j,FM^f}, \quad j = AS4, AS5, AS7, GT2$$

(29)

Finally, the probabilities associated to each functional status of the *NGCC* power plant are computed as stated in Eq. (30), considering their interrelations with the operative statuses of the gas and steam turbines as listed in Table 3 (where only those combinations which are feasible in accordance to Eqs. (22), (23),(24), (25), (26), (27), (28) and (29) are listed).

Table 3: Relation between binary variables for each section and functional status for the power plant

Binary variable associated to each section				Power plant functional status
GT1	GT2	ST1	ST2	
1	1	1	1	P1
1	1	1	0	P2
1	1	0	1	P2
1	1	0	0	P3
1	0	1	0	P4
0	1	0	1	P4
1	0	0	0	P5
0	1	0	0	P5
0	0	0	0	P6

$$Pr_{FS} = \sum_{FM^d} Pr_{FM^f}, \quad FM^d \subseteq FM^f$$

(30)

Component-Based Assessment of Maintenance Resources Impact

Allocation of extra resources on maintenance has a directly measurable effect on upholding or improving the repair time of a given piece of equipment, as a positive influence on the following aspects (among others) is observed:

- Fasten repairing equipment to acceptable standards
- Keeping inventory strategically, to ensure necessary materials are readily available
- Effectively applying manufacturers' recommendations and ensuring compliance with contractual requirements
- Maintenance staff training to improve their skills and capabilities

- Implementing methods to improve workplace security
- Systematizing maintenance actions and keeping personnel aware of applied policies
- Managing maintenance wastes

Enhancing any of these factors involves assigning extra resources for maintenance actions. As stated in Eq. (31), it is assumed that an exponential relationship exists between the mean time to repair of a piece of equipment and the maintenance funds assigned for such task: assigning more resources for equipment maintenance actions will improve its repair time, up to the point where technical difficulties constitute a speed limitation at which repairs can be performed (i.e. a point of diminishing return).

$$\mu AS = \mu_{AS,0} (FMant)^{\gamma AS} \tag{31}$$

The variables $\mu AS_{,0}$ and γAS depend upon the values of the minimum and maximum funds available for maintenance, and the associated values of the repair rates for such scenarios, as introduced in Eqs.(32) and (33).

$$\mu_{AS,0} = \frac{\mu_{AS,Min} - \mu_{AS,Max}}{(F_{Mant,Min})^{\gamma_{AS}} - (F_{Mant,Max})^{\gamma_{AS}}} \tag{32}$$

$$\gamma_{AS} = \frac{\ln \mu_{AS,Min} - \ln \mu_{AS,Max}}{\ln F_{Mant,Min} - \ln F_{Mant,Max}} \tag{33}$$

Parameters to be used in Eqs. (32) and (33)) can be computed from industry historic data on assigned maintenance funds versus achieved mean repair times ([NERC, 2011], [Participants, 2002] and [Alber et al., 1995]). Moreover, when the funds assigned for maintenance actions are increased from the minimum value *FMant Min* up to the maximum one *FMant Max*, it can be assumed that the repair rate per component improves by a given percentage *MFIRAS*, as introduced in Eq. (34) and listed in Table 4.

Table 4: Maintainability factor improvement rates

Item	Units	Value
Auxiliary services for the gas turbines (AS_1)	–	1.9
Gas turbine plus its associated generator (AS_2, AS_3)	–	1.5
Auxiliary services for both *HRSGs* and the steam turbine (AS_4)	–	1.9
Steam turbine and its associated generator (AS_5)	–	1.5
Heat recovery steam generators (AS_6, AS_7)	–	1.8

$$\mu_{AS,Max}=MFIRAS\mu_{AS,Min} \qquad (34)$$

Implementation

The mathematical program is implemented in the optimization software GAMS (Rosenthal, 2008) and solved through the algorithms CONOPT (Drud, 1996) and SBB (Drud, 2001). The proposed model comprises continuous and discrete variables, as well as highly non-linear constraints which configure a non-convex solutions space (including logarithmic mean temperature differences, correlations for water and steam properties according to [IAPWS, 1992] and [IAPWS, 2007], polytropic expansion at turbines, among others). Due to such characteristics, global optimal solutions cannot be guaranteed.

The initialization strategy of the optimization problem is outlined in Fig. 5, and implies:

- Common practical values are assigned to the power plant variables, when considering the design and operative characteristics of an actual one ([García and Mo nux, 2006], [Kehlhofer et al., 2009] and [Rapún Jiménez, 1999])
- The discrete variables associated to each component are set to 1 for the functional mode which represents nominal

operative capacity, while they are assumed as 0 at every other one

- For a given amount of resources assigned for maintenance actions, a preliminary evaluation of each component repair rate is achieved
- At this point, the economic performance of the project can be preliminarily computed, as if the system operated at full load across the whole time span.

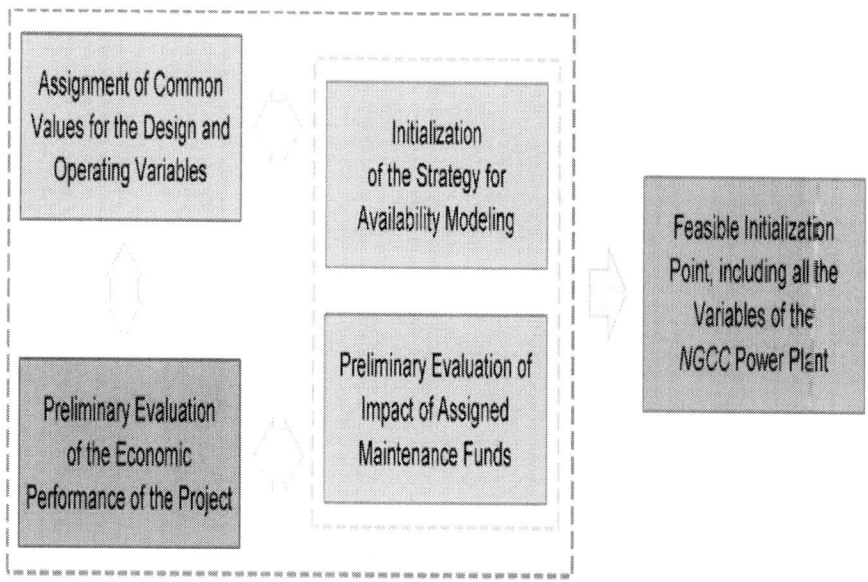

Figure 5: Initialization strategy.

This initial solution is then passed to the software GAMS, which starts the optimization procedure, and ultimately delivers optimal values for every continuous and discrete variable within the mathematical formulation, as summarized in Fig. 6 and including:

- Power plant variables effective fuel consumption and operative load of the gas turbine, design capacity and operative load of the steam turbine, exchange area and logarithmic temperature differences at each section of the *HRSGs*, pinch and approach points, boiler and cooling water consumption, pumps power requirement, characteristics (flow rate, temperature, pressure, composition) of each process stream

- Economic performance capital investment (including the main pieces of process equipment), operative expenditures (fuel, auxiliary services, maintenance, manpower, etc.), energy sales, total annual cost, electricity cost

- Availability modeling (according to Sections 2.4, 2.4.1, 2.4.2, 2.4.3, 2.4.4, 2.4.5, 2.4.6, 2.5 and 2.6): feasible functional modes (considering the value adopted for *NSIFC*), operative condition of each component at every feasible functional mode, transition rates (considering the value adopted for*NSE*), probabilities of functional modes, operative span for each functional mode, operative condition of each section, probabilities of functional statuses

- Evaluation of maintenance funds impact (according to Section 2.7): repair rate of each component as function of assigned maintenance funds.

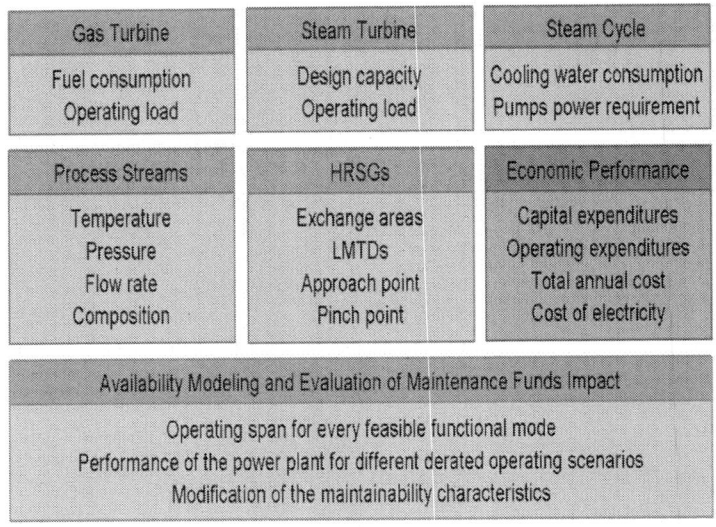

Figure 6: Decision variables for the power plant.

OPTIMAL DESIGNS: RESULTS AND DISCUSSION

Three case studies are hereafter solved and discussed, as summarized in Table 1 and briefly outlined below. As consequence, the economic optima of the *NGCC* power plant is obtained when solving the resultant *MINLP* formulation.

Case study 1 introduces the optimal design for the project while the availability and maintenance-related parameters are adopted at values usually adopted in the literature (as suggested byFrangopoulos and Dimopoulos (2004)). Results here obtained are compared against a *Reference plant* designed for a pre-specified annualized operative horizon and a fixed maintenance budget (seeGodoy et al. (2011)).

Case study 2 discusses the modifications of the optimal generation project as the availability-related parameters are varied across the whole range of feasible values; while the effect of varying the amount of resources assigned for maintenance actions is thoroughly analyzed in *Case study 3*.

In addition, a sensitivity analysis regarding the adopted economic parameters is presented, including fuel cost, investment on process equipment, as well as interest rate and life cycle span.

Table 1: Description of case studies

	Reference plant	*Case study 1*	*Case study 2*	*Case study 3*
Type of mathematical problem	*NLP*	*MINLP*	*MINLP*	*MINLP*
Strategy for availability modeling	Fixed annual operating horizon	State-space approach	State-space approach	State-space approach
NSE–NSIFC	N/A–N/A	1–1	1 to 4–1 to 7	*A*: 1–1

				B: 2–2
				C: 3–5
				D: 4–7
Assessment of maintenance funds impact	Fixed amount	Component-based policy	Component-based policy	Component-based policy
FMant	0.02	0.02	0.02	0.005–0.04

Case Study 1: The Simplest Implementation of the State-Space Approach

Optimal designs for the NGCC power plant are here obtained by solving the economic optimization formulation previously detailed in Section 2.2. *Case study 1* represents the simplest implementation of the state-space approach, as the availability and maintenance-related parameters are fixed at values usually adopted in the literature, which configure a back and forth radial feasible solutions region (i.e. where the system can only migrate from a fully operative state to one where at most a single component has failed, as exemplified in Fig. 3(a)).

Table 5 lists the probabilities of occurrence for every feasible functional status of the generation facility. A priori, when selecting $NSIFC = 1$ and $NSE = 1$, it is found that the model predicts a large time span for operative at full capacity, even with moderate values of the resources assigned for maintenance actions (adopted as 2% of the capital investment).

Table 5: *Case study 1*: probabilities of functional statuses

Functional status	Units	Value
P1	%	92.25
P2	%	2.15
P3	%	2.34
P4	%	2.39

P6	%	0.87

The obtained optimal values of the economic performance indicators of the project are presented in Table 6, and also compared against a *Reference plant* designed for a pre-specified annualized operative horizon and a fixed maintenance budget (obtained according to Godoy et al. (2011)).

Table 6: *Case study 1*: optimal economic indicators for the project

Item	Units	Reference plant	Case study 1
Total annual cost (*TAC*)	MUS$/y	339.07	349.14
Operative expenditures (*OPEX*)	MUS$/y	232.47	242.54
Capital expenditures (*CAPEX*/*CRF*)	MUS$/y	106.60	106.60
Generated energy (*GE*)	GWh/y	531.60	541.04
Energy sales (*Sales*)	MUS$/y	6.27	6.64
Cost of electricity (*COE*)	US$/MWh	54.06	52.54

It is observed that a 5.9% increment on the estimation of the total delivered energy, as the computation also includes the one generated at derated operative conditions. For similar reasons, the total annual cost increases by 3.0% as the operative expenditures are 4.3% larger (driven by the extra fuel consumed for electricity generation). On the other hand, the capital expenditures remain invariant since no relation is here considered between the equipment acquisition cost and the availability related parameters.

Therefore, it is here concluded that this first approach yields improvements over the evaluation of the project economics since the electricity cost diminishes by 2.8%, given that the delivered energy increases at a higher rate than the total annual cost.

Every component of the total expenses is disaggregated in Fig. 7. On an annual basis, the fuel consumption broadens 87.3% of the total raw material and utility costs, followed by the expenses on boiler and cooling water (9.0% and 3.7%, respectively).

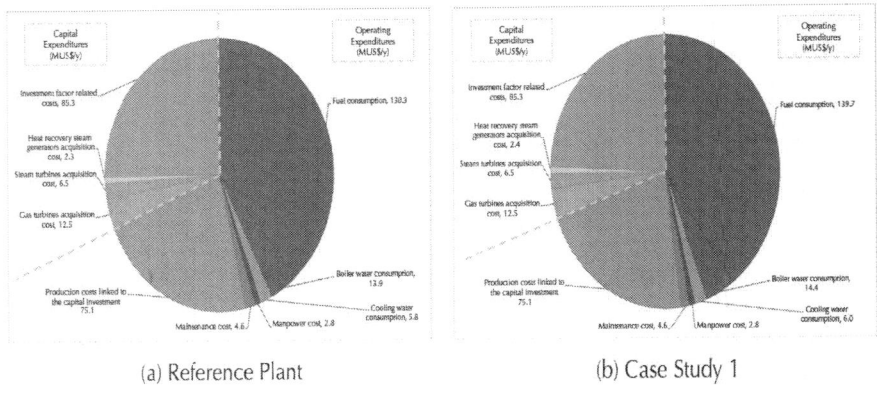

(a) Reference Plant　　　　　　　　(b) Case Study 1

Figure 7: *Case study 1*: costs distribution for the project.

The acquisition of the gas turbines requires 58.5% of the investment on process equipment, while the remaining 41.5% goes to the steam turbine and *HRSGs*. The construction of the facilities and other investment related factors take about 30% of the annualized expenditures.

In addition, Table 7 introduces the optimal values of key decision variables of the *NGCC* power plant, including design and operative ones.

Table 7: *Case study 1*: optimal design and operating variables for the power plant

	Units	Value
Power plant net generation capacity	MW	783.9
Gas turbine	MW	257.8
Steam turbine	MW	268.3
Thermal efficiency	–	0.5748
Gas turbine parameters		
Fuel flow rate	kmol/s	0.82

Compression ratio	–	15.8
Turbine inlet temperature	K	1560
Steam turbine flow rate		
Low pressure section	kg/s	98.3
Intermediate pressure section	kg/s	88.2
High pressure section	kg/s	68.8
HRSG exchange area		
Deaerator section	dam²	33.06
Low pressure section	dam²	35.80
Intermediate pressure section	dam²	43.94
High pressure section	dam²	80.54
HRSG operative pressure		
Deaerator section	MPa	0.152
Low pressure section	MPa	0.237
Intermediate pressure section	MPa	1.788
High pressure section	MPa	12.16
Reheater section	MPa	1.788
Utilities consumption		
Cooling water	kg/s	2847
Boiler water	kg/s	98.3

Note that the gas turbine design characteristics have been tuned to reproduce the performance of a commercially available one (GE PG9351FA). On the other hand, the steam cycle is specifically tailored for this application. Even so, optimal values for flow rates, temperatures, operative pressures, exchange areas, temperatures differences, etc. are in accordance with values previously reported in the literature ([Kotowicz and Bartela, 2010], [Srinivas, 2009], [Edris, 2010], [Woudstra et al., 2010] and [Franco and Giannini, 2006]).

Moreover, it is here noted that the operative variables of the NGCC power plant are allowed to adjust their values within wide ranges (as set by the selected minimum and maximum bounds on the technical constraints), which allows exploring a wider space of feasible solutions and enables attaining further improvement of the system performance.

Comparison with Other Authors

Frangopoulos and Dimopoulos (2004) introduced some simplifications when modeling the cogeneration plant (which are consistent with the ones at *Case study 1*) in order to obtain a formulation with a manageable size. In this context, they observed that an extra cogeneration package is necessary in order to satisfy the minimum availability requirements if reliability is taken into account, and proved that profits are overestimated when reliability aspects are ignored. In turn, El-Nashar (2008) utilized a single-transition structure to represent the feasible solutions space of a multistage desalination plant coupled to a cogeneration facility, which also allowed them to obtain a mathematical expression for the system availability. Then, they found that the water and power costs are higher when reliability considerations are considered during the design of the joint plant. Under similar modeling assumptions, improvement on economic indexes is here obtained, as no new equipment gets installed in order to increase the availability level of the system, while a more detailed evaluation of the project performance is achieved. It is also confirmed that the adoption of $NSIFC = 1$ and $NSE = 1$ generates a small and manageable optimization formulation.

Haghifam and Manbachi (2011) implemented a state-space and continued Markov model to study the reliability and availability of combined heat and power systems, where three subsystems are identified: electricity-generation, fuel-distribution and heat-generation. In turn, El-Nashar (2008) divided a cogeneration plant for power and desalination into three main components: gas turbine, *HRSG* and multistage flash plant. Likewise, Wang et al. (2013) proposed a multi-state model for a BHCP system while identifying three main supplies: electricity, heat and cold. Identification of key subsystems allowed these authors, and also in this work, to narrow down the number of logically arranged components (here listed in Eq. (2)), which is a reasonable assumption when analyzing the system as a whole during the synthesis and design stages of its life cycle.

Improving the reliability of each component implies attaining a higher level for the overall availability of the power plant, as demonstrated by Haghifam and Manbachi (2011). Meanwhile, repair rates are here improved as the necessary amount of maintenance resources are optimally assigned, while also evaluating their impact on the economic performance of the generation project. The parameters needed for the here proposed functionalities between repair rates and maintenance budget can be attained from industry historic data ([NERC, 2011], [Participants, 2002] and [Alber et al., 1995]); whereas other proposals ([Goel et al., 2002] and [Goel et al., 2003]) for assessing the impact of maintenance actions on the system availability would require not-easily-procurable data (which should be provided by equipment manufacturers in non-standard format).

When the operative status of each component is described by a zero-one variable, Vassiliadis and Pistikopoulos (2001) observed that the use of criteria that do not explicitly relate to process profitability as optimization objectives does not allow for the quantification of the balance between maintenance benefits and costs. This problem is here overcome as operative related financial flows (given by Eqs. (B.6), (B.7),(B.8), (B.9), (B.10) and (B.11)) are explicitly evaluated for every functional status when considering the associated optimal values of the operative variables (fuel consumption, auxiliary services, etc.), and afterwards weighted through Eq. (17).

Case Study 2: Influence of Availability Related Parameters NSIFC and NSE

Optimal designs for the *NGCC* power plant are here obtained by solving the economic optimization formulation previously detailed in Section 2.2. *Case study 2* analyzes the influence of modifications on the values adopted for the availability related parameters over the optimal economics of the generation project. Then, the number of simultaneously and independently failed components *NSIFC* is varied from 1 to 7, while the number of simultaneous events *NSE* is varied from 1 to 4. Consequently, the behavior of the system needs to be weighted across a broader span of feasible operative scenarios in order to evaluate its optimal economic performance.

Fig. 8 shows the main characteristics of the optimization formulation (as stated in Fig. 2) due to the modification of parameters *NSIFC* and *NSE*. Firstly, it is observed that the number of feasible functional modes (defined in Eq. (8)) grows exponentially along the value of *NSIFC* (see Fig. 8(a)), as does the number of continuous and discrete variables (see Fig. 8(b)), which is inherently linked to the size of the mathematical problem that needs to be solved.

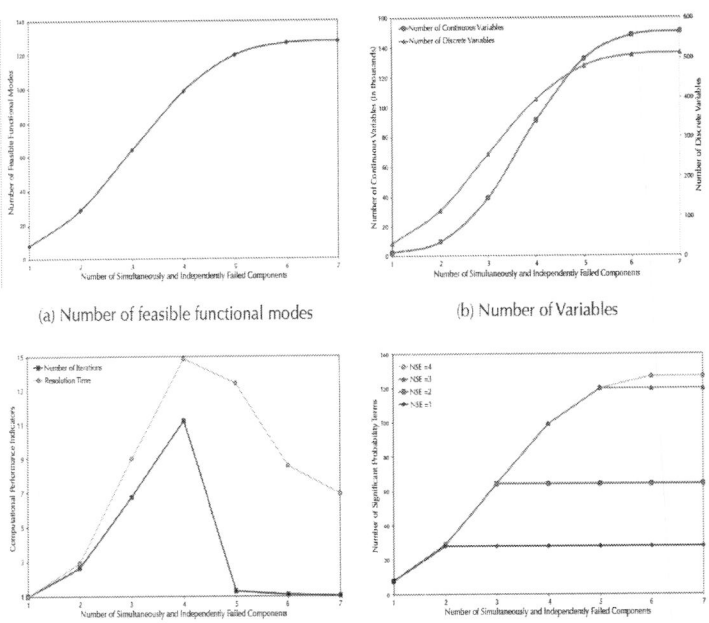

(a) Number of feasible functional modes   (b) Number of Variables

Figure 8: *Case study 2*: characteristics of the resultant mathematical problem for different *NSIFC* and *NSE* values.

Fig. 8(c) exhibits the computational cost for successfully achieving the resolution of each optimization problem, as the ratio against the resources needed for solving *Case study 1*. Note that a nested successive initialization strategy is used: for a given value of *NSE*, each optimal solution is used as starting point for the next problem where *NSIFC* is incremented in one unit; while *NSE* is increased by one unit in the outer loop. Even though, it is here observed that the resolution effort becomes several times larger than the requirements at the base case, thus imposing limitations to the selection of values for parameters *NSIFC* and *NSE* in order to efficiently solve the optimization problem.

For each pair of values of *NSIFC* and *NSE*, Fig. 8(d) shows the number of significant probability terms, i.e. the number of feasible functional modes which probability of occurrence is larger than 0.01%. For a given value of *NSE*, is here found that an increasingly larger number of feasible functional modes do not significantly contribute to the overall performance of the project as the value of *NSIFC* increases:

- For *NSE* = 1, the number of significant probability terms equals 8 for *NSIFC* = 1, and 28 for *NSIFC* ≥ 2
- For *NSE* = 2, the number of significant probability terms flattens at 64 for *NSIFC* ≥ 3. Even though, the number of feasible functional modes reach 128 for *NSIFC* = 7 (according to Eq. (8))
- Similar behavior is observed for *NSE* = 3 and *NSE* = 4

Then, it is concluded that some combinations of *NSIFC* and *NSE* do not significantly contribute to improving the accuracy on the evaluation of the economic performance of the generation project, even though they imply an increment on the required computational resources as previously discussed.

The probabilities of occurrence of each feasible functional mode for every optimization problem are introduced in Fig. 9. It is here observed that:

- When *NSIFC* = 1, the functional status *P5* cannot occur since it would require two components failing at the same time
- For *NSE* = 1, the probabilities remain almost invariant for the whole range of values of *NSIFC*
- As *NSE* adopts a value equal or greater than 2, the functional statuses which represent derated operative conditions

become increasingly more common, while the probability of operative at full capacity falls exponentially.

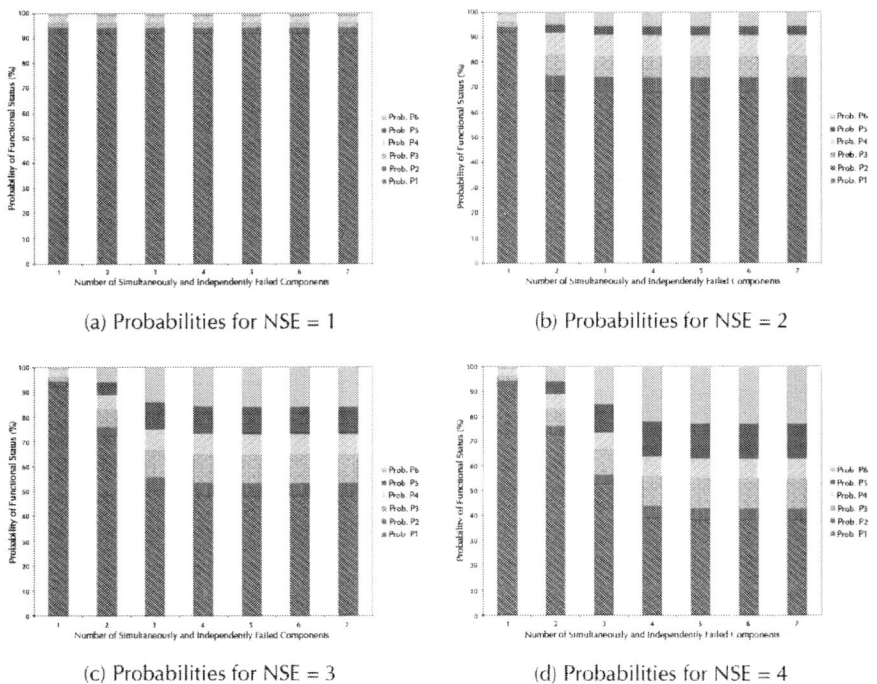

(a) Probabilities for NSE = 1

(b) Probabilities for NSE = 2

(c) Probabilities for NSE = 3

(d) Probabilities for NSE = 4

Figure 9: *Case study 2*: probability of functional statuses for different *NSIFC* and *NSE* values.

Fig. 10 presents the optimal economic indicators of the project for each pair of values for *NSIFC* and *NSE*. The total annual cost decreases as the value of *NSIFC* increases, for a given value of *NSE*, as shown inFig. 10(a). The main cost components exhibit similar trends, as illustrated for the fuel consumption in Fig. 10(b). Moreover, the total delivered energy also becomes smaller, as plotted in Fig. 10(c).

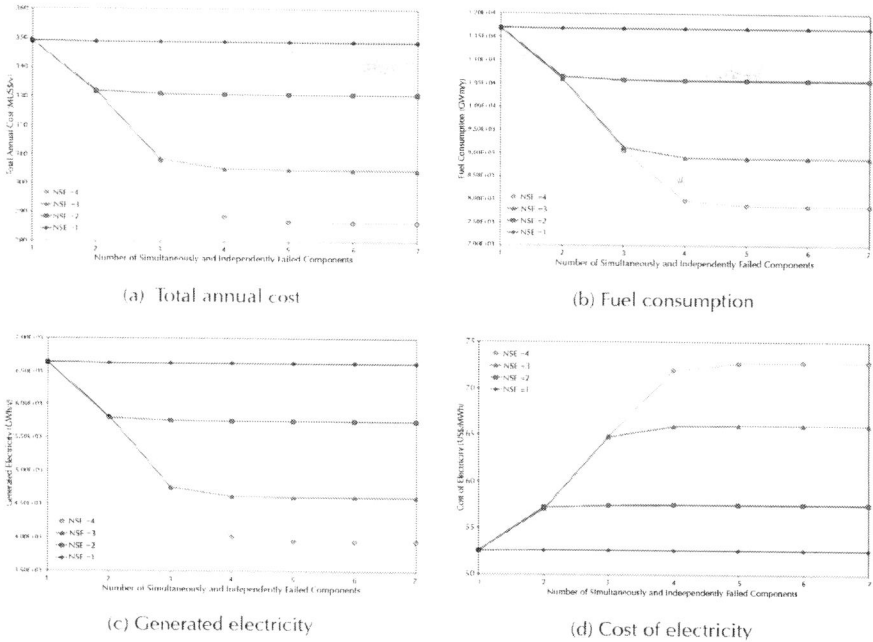

(a) Total annual cost

(b) Fuel consumption

(c) Generated electricity

(d) Cost of electricity

Figure 10: *Case study* 2: economic indicators for different *NSIFC* and *NSE* values.

Since the total annual cost decreases at a lower rate than the delivered energy's, it is observed in Fig. 10(d) that the cost of the generated electricity increases as the value of *NSIFC* does, for a given value of *NSE*. In addition, at first sight, the evolution of the objective function follows a similar trend that the one exhibited by the number of significant probability terms.

Thus, these trends regarding the evolution of the objective function can then be explained when considering the probability of occurrence of each feasible functional mode, since operative at a derated condition is more economically inefficient, and consequently implies a larger average (or weighted) cost per unit of delivered energy.

Case Study 3: Influence of Resources Assigned for Maintenance Actions

Optimal designs for the *NGCC* power plant are here obtained by solving the economic optimization formulation previously detailed in Section 2.2. *Case study 3* evaluates the modifications on the optimal performance of the generation project as the resources assigned for maintenance actions are varied from the minimum recommended budget (adopted as 0.5% of the capital investment), and up to the maximum available one (adopted as 4% of the capital investment).

Regarding the values of *NSIFC* and *NSE*, four feasible combinations are here adopted as listed in Table 1, and denoted as sub-cases *3.A, 3.B, 3.C* and *3.D*.

Fig. 11 introduces the evolution of the probability of occurrence of each functional status as the maintenance resources are varied. It is here observed that:

- The probability of operative at nominal capacity increases as the maintenance funds do, for every feasible scenario
- For *Case study 3.A*, which represents the simplest implementation of the state-space approach (i.e. when *NSIFC* = 1 and *NSE* = 1), the functional status probabilities exhibit only a slight dependence on the maintenance budget. Similar conclusions are drawn for those scenarios where *NSE* = 1 and independently of the value of *NSIFC*
- Increasing the maintenance resources has a positive and more pronounced effect over the economic performance of the generation project when *NSE* is equal or greater than 2, as shown for *Case studies 3.B, 3.C* and *3.D*.

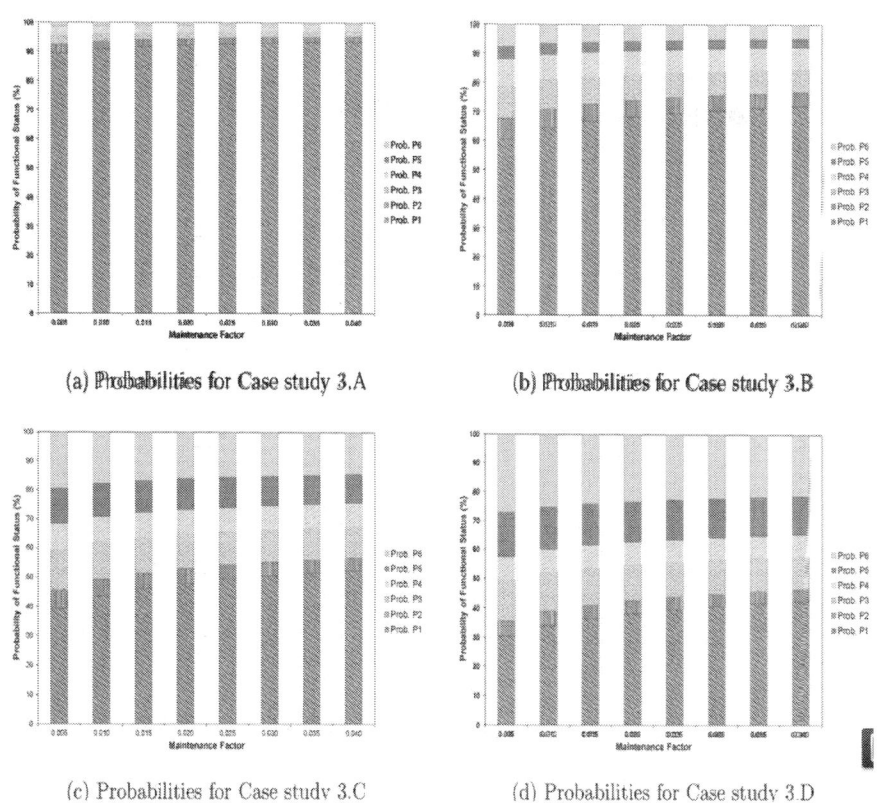

(a) Probabilities for Case study 3.A

(b) Probabilities for Case study 3.B

(c) Probabilities for Case study 3.C

(d) Probabilities for Case study 3.D

Figure 11: *Case study 3*: probability of functional statuses for different amount of resources assigned for maintenance actions.

Fig. 12 presents the optimal economic indicators for each sub-case within *Case study 3*, for the whole range of assigned maintenance resources.

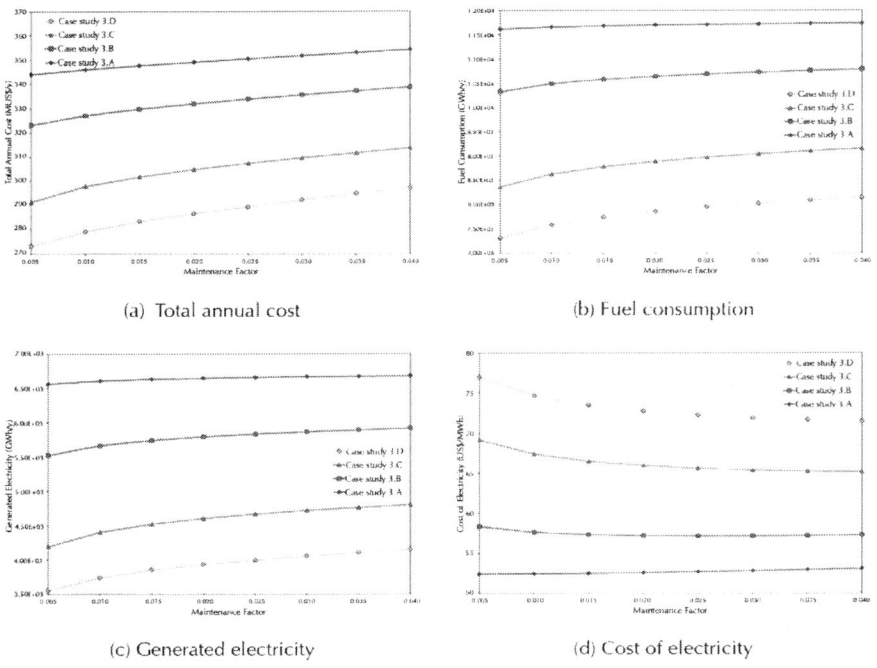

(a) Total annual cost

(b) Fuel consumption

(c) Generated electricity

(d) Cost of electricity

Figure 12: *Case study 3*: economic indicators for different amount of resources assigned for maintenance actions.

As an increasing maintenance budget implies a larger time span where the power plant operates at nominal capacity or at a derated condition with a higher operative load, the total annual expenditures present a positive slope, according to Fig. 12(a), since the main cost components also exhibit a similar trend, as exemplified in Fig. 12(b) for the fuel consumption. Consequently, an increment on the delivered energy is also attained, as Fig. 12(c) indicates.

Regarding the evolution of the electricity cost versus the assigned maintenance resources, Fig. 12(d) shows that the delivered electricity becomes cheaper as consequence of the positive effect exerted by an increasing maintenance budget.

Economic Sensitivity Analysis

As expected, the optimal values of the economic performance indicators of the project are critically dependent on the adopted values of the economic parameters. Thus, the sensitivity of the obtained optima is here discussed as several financial parameters are varied across a ±20% range, for each sub-case included in *Case study 3*, while the maintenance factor has been fixed at 0.020. Then, Fig. 13 reflects the relative influence of variations on the economic parameters over the electricity cost (i.e. the objective function).

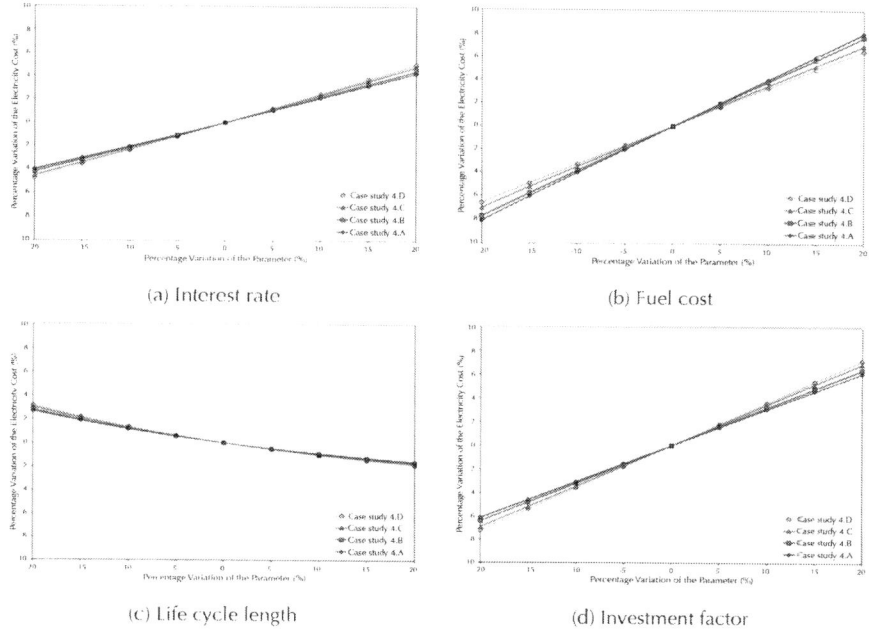

Figure 13: *Case study 3*: economic sensitivity analysis.

It is observed that the fuel cost exerts the largest negative impact on the electricity cost, followed in order of importance by the investment factor and the interest rate. On the contrary, increasing the life cycle length exposes a favorable (quasi) linear trend on the economic performance indicators of the project (as the capital expenditures get depreciated across a longer time span).

These economic parameters should then be carefully balanced considering the different available alternatives (turbines manufacturers, fuel sources, type of provision contract, etc.), so the newly designed generation plant results more appealing to potential investors.

Fig. 13 represents the economic sensitivity of the project as one economic parameter is varied at a time (while the other ones are kept at their expected values), which intends to configure a representative or average case. It is noted that the simultaneous increase of all the economic parameters (even including several others here not considered) would set a worst case scenario where the economic performance indicators get severely impacted and the electricity cost gets increased far beyond the values here reported; while a best case scenario could be obtained if the economic parameters are varied in the opposite direction, thus obtaining a minimum optimal value of the objective function.

A more rigorous and in-depth economic analysis should consider the uncertainty distribution (in a deterministic or stochastic way) of each economic parameter, which would enable finding the most likely scenarios that the project would have to face; although such analysis is beyond the scope of this work.

CONCLUSIONS

A comprehensive strategy has been here discussed in order to successfully achieve the economic optimization of a NGCC power plant, while considering the availability of the system through its wide array of feasible operative statuses, as well as the assignment of maintenance resources and its implications on the financial performance of the project.

Implications of a state-space approach are thoroughly discussed, where influence of maintenance funds on each component's repair rate is directly assessed. In this context, availability-related parameters *NSIFC* and *NSE* must be carefully selected in order to more accurately represent the actual characteristics of the region of economic optimal solutions for this type of generation plants, where it is observed that:

- A priori, utilizing *NSIFC* = 1 and *NSE* = 1, as illustrated in *Case study 1* and usually adopted in the literature, may render an improvement of the economic indexes. At *Case study 2*, similar conclusions are obtained when *NSIFC* = 1 for any value of *NSE*, or when *NSE* = 1 for any value of *NSIFC*
- When both *NSIFC* and *NSE* are equal or greater than 2, *Case study 2* shows that the power plant operates during an increasingly larger time span at different derated conditions, where the estimated value for the delivered energy cost results higher than for the *Reference plant*. It is also observed that functional status *P5* cannot be represented by the state-space method if *NSIFC* equals 1, since it requires two components failing independently at the same time
- An increasingly larger budget for maintenance actions improves the availability of the system, as quantified at *Case study 3*, whereas such behavior is more noticeable when both *NSIFC* and *NSE* are equal or greater than 2

Therefore, it is here found that values of parameters *NSIFC* and *NSE* are critical for the successful evaluation of the optimal economic performance of the generation plant. It is then observed that the value for both *NSIFC* and *NSE* should be carefully selected in order to better depict the actual characteristics of *NGCC* power plants across the full span of functional statuses under a component-based policy for assessment of the impact of assigned maintenance resources; although restraint should be exercised since the size of the resultant optimization mathematical problem becomes larger as *NSIFC* does, also requiring extra computational resources in order to promptly achieve convergence.

ACKNOWLEDGMENTS

The authors gratefully acknowledge the financial support of the Agencia Nacional de Promoción Científica y Tecnológica (ANPCyT), the Universidad Tecnológica Nacional (UTN) and the Consejo Nacional de Investigaciones Científicas y Técnicas (CONICET).

REFERENCES

1.   Abu-Zahra M, Niederer J, Feron P, Versteeg G. CO2 capture from power plants: Part II. A parametric study of the economical performance based on mono-ethanolamine. Int J Greenh Gas Control 2007];1(2): 135–42.

2.   Aguilar O, Kim J, Perry S, Smith R. Availability and reliability considerations in the design and optimisation of flexible utility systems. Chem Eng Sci 2008];63(14):3569–84.

3.   Alber T, Hunt R, Fogarty S, Wilson J. Idaho chemical processing plant failure rate database. Idaho Falls, USA: INEL; 1995].

4.   Bahadori A, Vuthaluru H. Estimation of performance of steam turbines using a simple predictive tool. Appl Therm Eng 2010];30(13):1832–8.

5.   Bassily A. Modeling, numerical optimization, and irreversibility reduction of a triple-pressure reheat combined cycle. Energy 2007];32(5): 778–94.

6.   Drud A. CONOPT: a system for large scale nonlinear optimization. Bagsvaerd, Denmark: ARKI Consulting and Development A/S; 1996].

7.   Drud A. SBB: a new solver for mixed integer nonlinear programming. Bagsvaerd, Denmark: ARKI Consulting and Development A/S; 2001].

8.   Edris M. Comparison between single-shaft and multi-shaft gas fired 800 MWel combined cycle power plant. Appl Therm Eng 2010];30(16):2339–46.

9.   El-Nashar A. Optimal design of a cogeneration plant for power and desalination taking equipment reliability into consideration. Desalination 2008];229(1–3):21–32.

10.   Erguina V. Safety assured financial evaluation of maintenance [Ph.D. thesis]. Texas A&M University; 2004].

11.   Franco A, Casarosa C. On some perspectives for increasing the efficiency of combined cycle power plants. Appl Therm Eng 2002];22(13):1501–18.

12.   Franco A, Giannini N. A general method for the optimum design of heat recovery steam generators. Energy 2006];31(15):3342–61.

13.   Frangopoulos C, Dimopoulos G. Effect of reliability considerations on the optimal synthesis, design and operation of a cogeneration system. Energy 2004];29(3):309–29.

14.   García S, Mo nux F. Centrales térmicas de ciclo combinado: teorí a y práctica. Ediciones Díaz de Santos; 2006].

15.   Gjorgiev B, Kancev ˘ D, Cepin ˘ M. A new model for optimal generation scheduling of power systemconsidering generation units availability.IntJ Electr Power Energy Syst 2013];47 129–39.

16.   Godoy E, Benz S, Scenna N. A strategy for the economic optimization of combined cycle gas turbine power plants by taking advantage of useful thermodynamic relationships. Appl Therm Eng 2011];31:852–71.

17.   Godoy E, Scenna N, Benz S. Families of optimal thermodynamic solutions for combined cycle gas turbine (CCGT) power plants. Appl Therm Eng 2010];30(6/7):569–76.

18.   Goel H, Grievink J, Herder P, Weijnen M. Integrating reliability optimization into chemical process synthesis. Reliab Eng Syst Saf 2002];78(3): 247–58.

19.   Goel H, Grievink J, Herder P, Weijnen M. Optimal reliability design of process systems at the conceptual stage of design. In: Annual reliability and maintainability symposium, 2003. IEEE; 2003]. p. 40–5.

20. Haghifam M, Manbachi M. Reliability and availability modelling of combined heat and power (CHP) systems. Int J Electr Power 2011];33(3):385–93.

21. Henao C. Simulación y evaluación de procesos químicos. Medellín, Colombia: Universidad Pontificia Bolivariana; 2005].

22. IAPWS. Revised supplementary release on saturation properties of ordinary water substance. St. Petersburg, Russia: The International Association for the Properties of Water and Steam; 1992].

23. IAPWS. Revised release on the IAPWS industrial formulation 1997 for the thermodynamic properties of water and steam. Lucerne, Switzerland: The International Association for the Properties of Water and Steam; 2007].

24. Ibe O. Markov processes for stochastic modeling. Elsevier Academic Press; 2009]. Iyer R, Grossmann I. Optimal multiperiod operational planning for utility systems. Comput Chem Eng 1997];21(8):787–800.

25. Kehlhofer R, Rukes B, Hannemann F, Stirnimann F. Combined-cycle gas & steam turbine power plants. 3rd ed. Pennwell Books; 2009].

26. Kotowicz J, Bartela L. The influence of economic parameters on the optimal values of the design variables of a combined cycle plant. Energy 2010];35(2): 911–9.

27. KuoW, Zuo M. Optimal reliability modeling: principles and applications. New Jersey, USA: Wiley; 2003].

28. Lisnianski A, Elmakias D, Laredo D, Ben Haim H. A multi-state Markov model for a short-term reliability analysis of a power generating unit. Reliab Eng Syst Saf 2012];98(1):1–6.

29. Luo X, Zhang B, Chen Y, Mo S. Operational planning optimization of steam power plants considering equipment failure in petrochemical complex. Appl Energy 2013];112:1247–64.

30. Martelli E, Amaldi E, Consonni S. Numerical optimization of heat recovery steam cycles: mathematical model, two-

stage algorithm and applications. Comput Chem Eng 2011];35(12):2799–823. Matches. Mathematics and chemistry; 2013] http://www.matche.com

31. Mc Leod J, Rivera S, Barón J. Optimizing designs based on risk approach. In: Proceedings of the World Congress on Engineering, vol. 2. Citeseer; 2007]. p. 1044–9.

32. NERC. 2006–2010 generating availability report. New Jersey, USA: NERC; 2011].

33. Nye Thermodynamics Corporation. Gas turbine prices; 2013] http://www.gasturbines.com

34. OREDAParticipants. Offshore reliability data handbook. 4th ed. Trondheim, Norway: OREDA; 2002]. Rao A, Rubin E. A technical, economic, and environmental assessment of aminebased CO_2 capture technology for power plant greenhouse gas control. Environ Sci Technol 2002];36(20):4467–75.

35. Rapún Jiménez J. Modelo matemático del comportamiento de ciclos combinados de turbinas de gas y vapor [Ph.D. thesis]. Spain: ETSII-UPM; 1999] [in Spanish].

36. Rosenthal R. GAMS: a user's guide. Washington, DC, USA: GAMS Development Corp; 2008].

37. Seider W, Seader J, Lewin D. Product & process design principles: synthesis, analysis and evaluation. USA: Wiley & Sons, Inc; 2009].

38. Srinivas T. Study of a deaerator location in triple-pressure reheat combined power cycle. Energy 2009];34(9):1364–71.

39. Sun L, Liu C. Reliable and flexible steam and power system design. Applied Thermal Engineering; 2014].

40. Tan J, Kramer M. A general framework for preventive maintenance optimization in chemical process operations. Comp Chem Eng 1997];21(12):1451–69.

41. Terrazas-Moreno S, Grossmann I, Wassick J, Bury S. Optimal design of reliable integrated chemical production sites. Comput Chem Eng 2010];34(12):1919–36.

42. Ulrich G, Vasudevan P. How to estimate utility costs. Chem Eng 2006];113(4): 66–9.

43. U.S. Department of Energy. U.S. Energy Information Administration; 2013] http://www.eia.gov

44. U.S. Energy Information Administration. Updated capital cost estimates for electricity generation plants. Washington, DC, USA: U.S. Department of Energy; 2010].

45. Vassiliadis C, Pistikopoulos E. Maintenance scheduling and process optimization under uncertainty. Comp Chem Eng 2001];25(2/3):217–36.

46. Wang J-J, Fu C, Yang K, Zhang X-T, Shi G-h, Zhai J. Reliability and availability analysis of redundant BCHP (building cooling, heating and power) system. Energy 2013];61:531–40.

47. Woudstra N, Woudstra T, Pirone A, Stelt T. Thermodynamic evaluation of combined cycle plants. Energy Convers Manage 2010];51(5):1099–110.

Availability and Reliability Considerations in the Design and Optimisation of Flexible Utility Systems

Oscar Aguilar, Jin-Kuk Kim, Simon Perry, and Robin Smith

Centre for Process Integration, School of Chemical Engineering and Analytical Science, The University of Manchester, PO Box 88, Manchester, M60 1QD, UK

ABSTRACT

Industrial utility plants are usually comprised of many interconnected units that must constitute a flexible and reliable system capable of meeting process energy requirements under different circumstances (e.g. varying prices, demands, or equipment shutdowns). Also, in order to avoid large economic penalties, the design and operation of a utility plant should consider that the equipment is not fully reliable and that each item needs to receive preventive and corrective maintenance. Conventionally, these issues are handled by installing additional units according to rules of thumb or heuristics, which usually imply excessive capital costs and might even result in designs that cannot satisfy the specified demands for certain situations. In contrast, during the present work a systematic methodology has been developed to address the design and operation of flexible utility plants incorporating reliability and availability considerations. The suggested method is based on a novel modelling and optimisation framework that can address grassroots design, retrofit, or (pure) operation problems in which design and operational parameters are optimised simultaneously throughout several scenarios. Thereafter, it is possible to define maintenance and failure situations in different operating periods to ensure that the plant will be able to cope with them, while meeting process requirements at minimum cost. Hence, for design cases, the most cost-effective elements of redundancy can be determined without pre-specifying any structural options in the final configuration (e.g. equipment sizes, types, and number of units). Furthermore, the proposed (multiperiod) MILP formulation is robust enough to tackle problems of the size and complexity commonly found in industry, and has the potential of yielding significant economic savings.

INTRODUCTION

The design and operation of industrial utility systems offer multiple degrees of freedom (e.g. equipment sizes, number of units, and their loads) that can be exploited to achieve large economic savings. At the same time, since no equipment is one hundred percent reliable, it is necessary to account for the possibility of failure together with preventive and corrective maintenance periods of all the items of the plant.

In this way, reliability and availability issues not only have a major impact on the design and operation of a utility system, but also they considerably increase the number of options that should be assessed to reduce capital and/or operating costs. Hence, minimising such expenditure represents a very challenging task due to the highly combinatorial computations involved and strong interrelations between the equipment. This means that the whole utility system must be simulated in order to take into account all the potential units and determine the plant-wide consequences of any proposed modification. Moreover, only with this approach it is possible to assess which design and/or operational decisions actually improves the overall economics of the system.

Consequently, mathematical programming techniques have been broadly applied to study the design and operation of utility systems as optimisation tasks that search for the options minimising a given objective (e.g. to reduce costs), while meeting the specified demands and other practical constraints. It is possible to identify previous pieces of research dealing only with operational problems for existing plants in which structural modifications are not considered (e.g. Nath and Holliday, 1985, Iyer and Grossmann, 1997 and Ashok and Banerjee, 2003). Other approaches have addressed the design of utility systems assuming that all units operate at full load to satisfy a single set of (constant) demands and conditions (e.g.Papoulias and Grossman, 1983 and Bruno et al., 1998). Due to the limitations of these types of studies, there have been methodologies combining elements of both operational and design problems (e.g. Hui and Natori, 1996, Maia and Qassim, 1997, Maréchal and Kalitventzeff, 2003 and Varbanov et al., 2005), but without simultaneously optimising equipment sizes and loads as continuous functions (i.e., requiring pre-specified design options that significantly limit the solution space of the problems) and/or involving complex optimisation procedures with limited applicability to real cases. More recently, Aguilar et al. (2007a) proposed a robust computational tool to address grassroots design, retrofit and operational problems for flexible utility plants, considering structural and operational parameters as variables to be optimised at the same time.

Regarding availability and reliability issues, due to the complexities of handling these concerns within a systematic methodology, in most cases the redundant elements of a utility plant are pre-specified according to heuristics or rules of thumb. For instance, once a promising design (without redundancy) has been obtained by some means, it might be decided to install an arbitrary number of additional units to perform the same function and/or to oversize the equipment by a certain factor. A popular heuristic of this type is the n+2 rule where two extra units are added to an initial design in which redundancy has not been included. Similar heuristics have been incorporated into some methods employing mathematical programming techniques. For example, in the work by Varbanov et al. (2004), the sum of all boiler sizes is designed to be 30% higher than the steam requirements. Nevertheless, in these cases there are numerous degrees of freedom that are not being exploited and which would prevent the design from incurring in excessive capital costs. Moreover, since no failure or maintenance scenarios are actually checked, it is possible that the final system will not be able to fulfill the expected demands even during a simple maintenance condition (i.e., the extra capital costs are not strategically invested).

There have been other approaches employing computational routines to address the reliability and availability of energy systems without pre-defining any elements of redundancy. For example, Olsommer et al., 1999a and Olsommer et al., 1999b proposed a methodology to optimise the design and operation of a cogeneration plant subject to different operating scenarios. The approach also includes an automated program to perform a reliability analysis and determine all the possible failure modes of the optimal system (obtained in first place). Thereafter, the penalties from the scenarios in which the proposed design cannot fully meet the demands are calculated (i.e., charges are applied per unit of heat or power below the stipulated requirements) and, with this information, the overall operating cost of the plant is rectified. In this way, although reliability issues are considered after the optimisation, it is possible to compare different configurations and obtain more realistic operational costs. This method was extended by Frangopoulos and Dimopoulos (2004) who used similar steps, but all of them within an optimisation procedure. In this case, a search routine based on a genetic algorithm first proposes a design for which reliability analysis is carried out, generating all possible failure scenarios (for each original operating period) and calculating the penalties whenever the demands are not fully satisfied. The corresponding investment and overall operating costs are then compared with many other design proposals automatically generated by the routine, until a near-optimum solution is considered to be found.

The major drawback of the two methodologies just mentioned is large computational resources needed to solve the optimisation tasks, which limit both formulations to cases with a restricted number of design alternatives and/or operating scenarios. On the other hand, Del Nogal et al. (2005) employed non-linear (MINLP) programming techniques for the design of utility plants for power-intensive processes. In this method, the expected availabilities (i.e., down-time fractions) for different equipment types, sizes, and configurations are directly included in the optimisation procedure so that capital costs, operating costs, and profit losses (due to overall plant unavailability) are considered simultaneously. Although the solving routine is robust enough to deal with practical problems, the formulation only handles a single operating scenario assuming average conditions.

In contrast, the present work has extended the methodology proposed by Aguilar et al. (2007b) to address the design and operation of flexible utility systems by incorporating reliability and availability issues. Apart from grassroots design cases, the suggested multiperiod (MILP) optimisation framework is robust enough to tackle retrofit and pure operational problems of the size and complexity commonly found in industry. Also, the approach determines the most convenient maintenance schedule for the equipment during the whole time horizon, and is capable of considering failure situations for all scenarios. In this way, large capital savings can be obtained since the optimal redundancy for the system is established taking into account normal, maintenance and/or failure conditions without pre-defining any redundant elements in the final plant configuration. Furthermore, users have the possibility of including failure situations that they consider pertinent and compare the results for different circumstances.

BASIC DEFINITIONS

Reliability can be defined (Ebeling, 1997) as the probability that a device or system will perform a required function at a given point in time, when operated under specific conditions. In other words, reliability is a quantitative measure (e.g. percentage) of non-failure operation over an (operational) time interval. It is important to note that this definition assumes that certain criteria have been previously established to clearly specify what is considered to be the intended function of the item. For instance, a utility plant that is importing electricity from the public grid due to the failure of a generator would be still complying with its duty of distributing power to other processes, even though it is using an emergency connection. Also, it is essential to indicate the application and operating conditions under which an item will be employed. For example, a gas turbine driving an electric generator at rated load would exhibit higher reliability than a unit running a compressor subject to frequent load variations. In addition, reliability should be also specified for a given period of time (in which a unit is operating) as this variable also has an important effect. For instance, a compressor becomes less reliable as its number of operating hours (without being switched off for maintenance) increases.

In addition, failure normally implies a corrective maintenance action corresponding to the repair time needed to bring an element back to regular operation. Also, the equipment often requires preventive or scheduled maintenance to improve its reliability. Hence, the down time of a unit is comprised of both its corrective and preventive maintenance periods. Thereafter, *availability* is defined as the probability that an item performs its required function during a certain interval of time (including maintenance periods) whilst operated under specific conditions. In other words, it is the percentage the item is operating over a specified time interval encompassing also its maintenance periods.

It is worth mentioning the concept of *maintainability* which might correspond to the probability of completing a repair action during a certain time interval (i.e., how quickly an item can be repaired due to failure; Olsommer et al., 1999a and Olsommer et al., 1999b), or it can denote the probability of performing preventive maintenance during a certain time interval (i.e., how often the maintenance action should be;Govil, 1983). In any case, it must not be confused with the downtime or with the complement of availability (i.e., one minus availability). Thus, while reliability is a measure of non-failure operation of an item, maintainability is related to its capability of being repaired or to its need of receiving maintenance. Thereafter, both measures can be used to calculate the overall availability, representing the net operational time of an element or system.

One of the most common ways of coping with these issues is to resort to some kind of *redundancy*, which can be defined as the strategy of employing more than one element functionally connected in parallel (i.e., to perform the same action) to achieve higher reliability. Note that units or elements of different types might be installed to carry out a common function. For instance, during the failure of a heat recovery steam generator, an oil-fired boiler might be providing the required steam to satisfy the specified demands. Furthermore, the additional (i.e., redundant) components might be performing either an *active redundancy*if they are normally operating together with other units, or a *passive redundancy* if they are ready to operate during a failure condition, but are normally switched off (see Fig. 1).

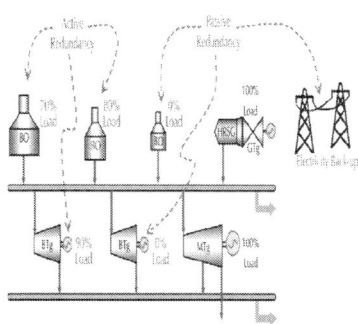

Figure 1: Active and passive redundant elements within a utility system.

REDUNDANCY ISSUES IN UTILITY SYSTEMS

Since utility plants must be flexible and reliable to cope with many circumstances, they are usually comprised of numerous elements capable of providing different types of redundancy in order to satisfy the required demands. The situations under which redundancy is needed can be grouped into four categories (i.e., sources of redundancy):

- *Demands variations*: Given that the utility plant is subject to fluctuating requirements, it must be able to fulfill the expected consumption peaks from external processes, even if they occur sporadically.

- *Scheduled equipment maintenance*: Whenever a piece of equipment within the utility plant is switched off to receive preventive maintenance, other elements of the system must compensate for the unavailability of one or more units.

- *Equipment failure*: Every time a unit of the utility plant must be shut down due to (unexpected) failure, the system should have enough redundancy to ensure that the demands will be met even if other pieces of equipment are down for maintenance and/or repair.

- *Special operating conditions*: This category includes transient situations affecting equipment performance along with other special circumstances, such as plant or equipment startups and shutdowns. For example, gas turbines might be oversized to deal with high ambient temperatures, which diminish their power output and efficiency during the summer season.

If a utility system cannot deal with such conditions, at some point it will not be able to fully satisfy the specified requirements. In those cases, there will be additional costs due to contractual penalties or from production and profit losses, which usually dominate the economics of the plant and should be avoided. Therefore, as can be appreciated in Fig. 1, these systems are normally provided with several redundant components that might operate at partial loads (i.e., active redundancy), or might be normally switched off (i.e., passive redundancy). Emergency back-up from the electricity grid also represents an element of passive redundancy. In fact, deciding the appropriate redundancy elements and how they should be operated during a given situation are complex tasks involving many issues as explained next.

Operation and Maintenance of Redundant Components

In the case of existing utility systems where structural modifications are not contemplated, the elements of redundancy have been already provided and cannot be modified. However, reliability and availability issues will still affect plant operation significantly. As explained in Aguilar et al. (2007b), the operational problem of a utility system consists of finding the operating conditions of the whole plant through each scenario that minimise/maximise an objective defined for the task (e.g. reducing overall costs). In this way, once reliability and availability concerns are incorporated, the problem becomes more complicated since it is necessary to consider not only normal operating circumstances (i.e., when all units are available), but failure and maintenance situations also.

Consequently, for long- or mid-term cases (e.g. a year or more), it is required to establish the maintenance scheduling for all the equipment in the utility plant given a set of expected demands and site conditions over several operating periods. Furthermore, it is possible for the units to fail in any scenario, including periods when other pieces of equipment are down for maintenance, and/or failure. Thus, when evaluating any of such cases (normal, maintenance, and/or failure situations) the operating conditions (e.g. on/off status and load) for all the units must be determined in case it is feasible to meet the demands. This, in turn, will define not only if there is enough redundancy to cope with such circumstances, but also the specific type of redundancy that every element should perform (i.e., active or passive) in different scenarios. Additionally, it is also necessary to take into account if the redundant components would be able to respond quickly enough to an unexpected failure situation. For example, it might not be possible to start a boiler in a few minutes to compensate for a steam generator that suddenly is forced to shut down, whereas boilers already in operation might be able to increase their loads almost immediately.

Moreover, due to the large number of connections between units, the degrees of freedom corresponding to all the elements of the system (including redundant ones) are highly interrelated and modifying the operation of any item would have an impact on the rest of the plant. As a result, reliability and availability issues should be considered in an integrated approach to assess their plant-wide consequences and verify whether a given decision can improve the overall economics of the utility system.

Redundancy during the Design of Utility Systems

For retrofit or grassroots design problems it is possible to modify or to fully decide the redundant elements to be provided. As explained by Aguilar et al. (2007b), in these cases the objective is to determine the types and sizes of new units to be installed (including redundant ones), their number and interconnections, together with the operational conditions of the whole plant (in each scenario) that will minimise/maximise an objective defined for the task (e.g. reduce total costs). Again, design problems become even more complicated when reliability and availability issues are included, not only due to the additional operating situations to be considered (i.e., normal, maintenance, and failure), but also because of the larger number of units that must be installed to cope with them. Furthermore, the multiple design and operational degrees of freedom are highly interrelated all together and complex calculations are also involved during assessment different alternatives.

The interactions between design and operating variables are particularly evident when reliability and availability concerns are taken into account. For instance, it might be decided to employ a large boiler intended for part-load operation during most of the time (i.e., active redundancy) or to install several smaller units, one of which would normally be switched off (i.e., passive redundancy). Also, it should be considered that, while a large piece of equipment would be more efficient and less expensive than various small ones (summing up the same capacity), several units in parallel might offer a better reliability (i.e., if any of them fails there would be more to compensate it). As can be appreciated, in order to determine the most convenient design of the plant it is necessary to consider how the units to be installed would operate under normal conditions and during maintenance and failure scenarios.

Fig. 2, Fig. 3 and Fig. 4 illustrate how an initial design for a utility system might be modified to provide redundant elements and deal with reliability and availability concerns. Without consideration of redundancy, a utility system does not need to accommodate redundancy as shown in Fig. 2 (i.e., all the equipment is based on full load.). As shown on Fig. 3, redundancy can be introduced in several ways: (1) installing more units of the same type, (2) installing more units of different types but performing the same function, (3) increasing equipment sizes of any type, (4) importing more utilities, and/or (5) exporting less utilities. For instance, while only two boilers were contemplated in the original design, the final configuration includes another boiler (i.e., same type) and a heat recovery unit that is also producing high-pressure steam (i.e., different equipment type with the same function). Similarly, another steam turbine, a gas turbine (both driving electric generators), and an emergency connection with the external grid ensure a reliable electricity supply. Fig. 4 illustrates why redundant capacity or equipment is required for the utility system, and how appropriate actions (increasing and distributing working loads of boiler and steam turbine according to failure conditions) are taken.

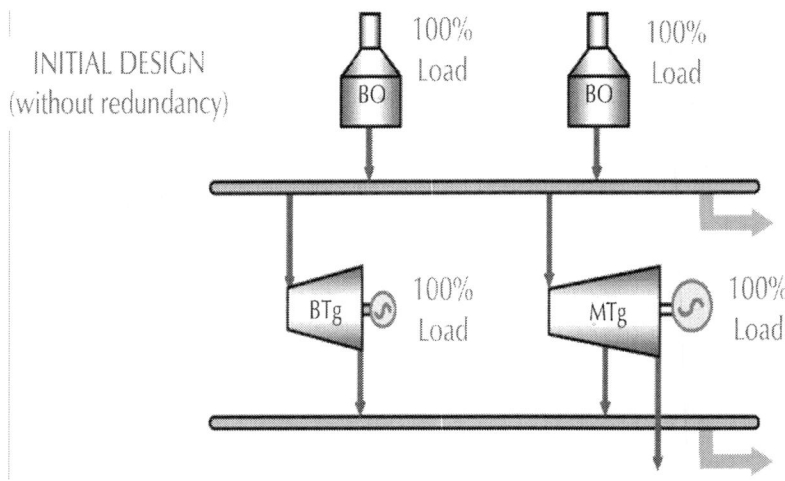

Figure 2: A utility system without elements of redundancy.

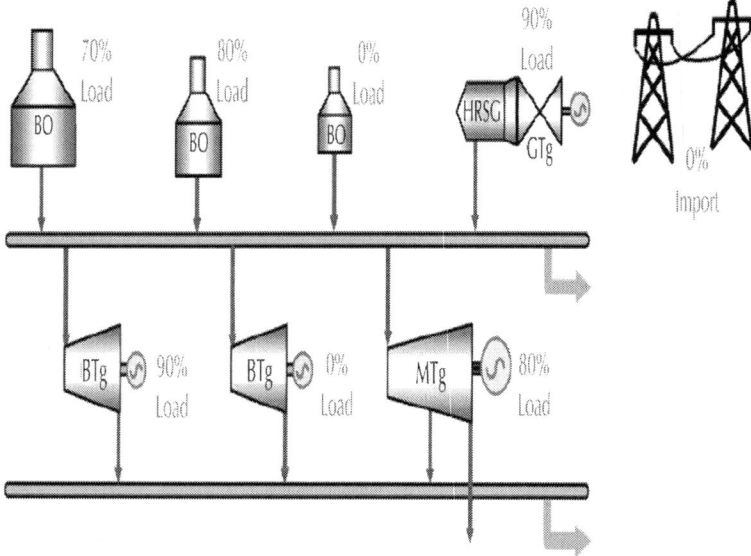

Figure 3: A utility system with elements of redundancy operating under normal conditions.

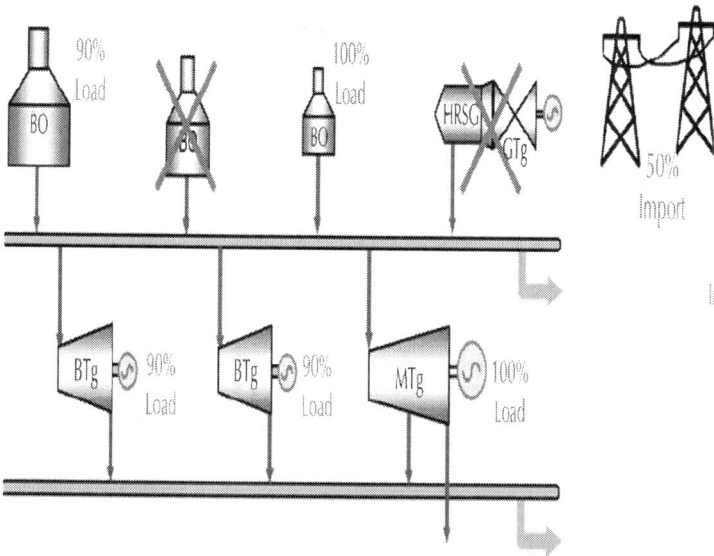

Figure 4: A utility system with elements of redundancy operating under a failure condition.

From these examples it is evident that there are many degrees of freedom to take into account while defining redundant elements (and their operation) within a utility plant. Moreover, due to the high interrelation between all the components of the system, these variables cannot be isolated from the rest of the problem. On the contrary, they should be addressed with an integrated approach for the design and operation of utility systems, incorporating reliability and availability issues.

DESIGN AND OPTIMISATION OF UTILITY SYSTEMS WITH REDUNDANCY

As mentioned in the introduction, an important characteristic of previous approaches using mathematical optimisation procedures to deal with the availability and reliability of energy systems is that they rely on the trade-off between investment and penalty costs for not fulfilling the specified requirements. Although for some particular problems such penalties might be calculated accurately, that is not the case in most situations. For instance, if an energy system is supplying the needs of a single processing plant, it might be possible to assume that all the utility demands are proportional to the production rate. Hence, profit losses due to unavailable equipment could be approximated using the product price as a fixed reference (e.g. overall revenue proportional to production). A similar approach could be applied to a cogeneration plant that is selling utilities at specific (i.e., contractual) prices.

Nevertheless, industrial systems usually must deliver different types of utilities to several (internal) processes, which are not necessarily related to each other. Hence, there is not a clear basis to estimate the economic consequences of the energy requirements that are not fully satisfied. Moreover, production deliveries are the economic priority in most of those sites and the penalties for not meeting the expected demands would be very high. Therefore, a more practical approach for these cases is to trade off capital costs against different failure situations over which the stipulated set of utilities must still be supplied. For example, a system with enough redundancy to handle the simultaneous shutdown of three pieces of equipment (and yet meeting the demands) would be more expensive than one that could only manage two units down. Consequently, during the proposed methodology it is assumed that the prescribed requirements in every operating period must be met by the utility plant. Thereafter, it is necessary to specify (apart from the normal conditions when all the equipment is available) the maintenance and/or failure scenarios that should be considered to ensure that the system would be provided with the adequate redundant elements to cope with them and minimise costs.

Even though it is not viable to enumerate all the possible failure modes for a typical utility plant (for a system of 10 elements the number of mathematical failure modes is $2^{10}=1024$), the probability of incidence for most of these scenarios is very low (e.g. the simultaneous failure of four units out of six) and a few of them can be selected in order to provide a practical solution. Thus, since the suggested optimisation framework is a multi-period one, users have the possibility of setting up many situations and compare the results for different (realistic) circumstances. Furthermore, given that the redundant pieces of equipment of the same type (obtained with the optimisation) are often of the same size, the practical failure and/or maintenance conditions can be represented with even fewer scenarios. Note that these results from the optimiser confirm the engineering practice of distributing evenly the reliability among several units (i.e., do not rely more on any of them).

Sets of Time Periods

A basic element for addressing reliability and availability issues within complex systems is the specification of various scenarios to represent different (normal, maintenance and/or failure) operating conditions. As illustrated on Fig. 5, in the suggested methodology it is possible to define seasonal and inter-seasonal time periods so that for each interval of the first kind there is an associated set of periods of the second type. Therefore, all intervals and their corresponding time fractions are, in fact, identified by two indexes (t and θ) as expressed in Eq. (1) and Eq. (2). Note that the duration of each period is equal to its time fraction multiplied by the total number of hours in the time horizon and, thus, the added fractions must yield one. In this way, it is easier to characterise certain parameters such as electricity prices, which can vary on an hourly basis, but that can also exhibit changes in their value, number, and distribution for different seasons (e.g. peak and off-peak hours for summer; but peak, semi-peak, and off-peak for winter). Furthermore, from a mathematical point of view this distinction allows establishing optimisation constraints between inter-seasonal periods, and/or across seasons. Such a feature is particularly useful when dealing with the operation of redundant equipment as described next.

$$Time_{t,\theta} = f_{t,\theta} \cdot hrs^{tot}$$

(1)

$$\sum_{t}\sum_{\theta} f_{t,\theta} = 1$$

(2)

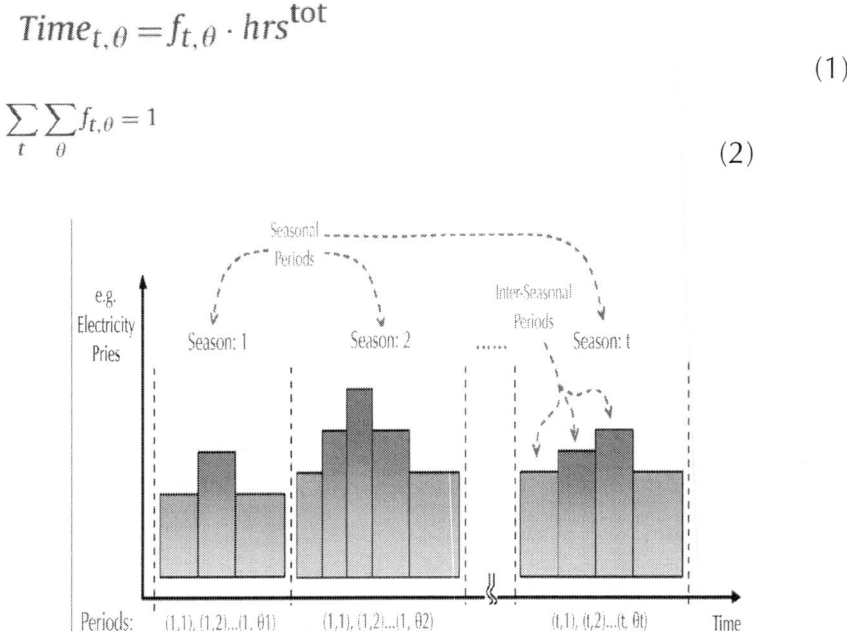

Figure 5: Different types of operating scenarios can be defined to address reliability and availability issues with the proposed methodology.

Scheduling and Redundancy for Equipment Maintenance

As discussed previously, even for existing utility plants where design changes are not contemplated, reliability and availability issues must still be taken into account while optimising the operation of such systems. Therefore, whenever an operational problem is posed as an optimisation task, there are additional constraints dealing with those concerns that should be incorporated into the mathematical formulation. The first of these restrictions is Eq. (3), which forces every piece of equipment in the system to be switched off for a minimum number of seasonal periods throughout the time horizon, so that the units can receive preventive maintenance. The time horizon should be large enough and the seasonal periods equal or smaller than the maintenance intervals. For example, all the boilers of a plant might be required to shut down for one month per year or two months in five years. It is important to note that, since the maintenance periods are not pre-defined in any way, the optimisation determines the most convenient scheduling for each unit considering all the scenarios originally specified.

Another concern mentioned before is whether a given piece of equipment would be ready to operate if it is suddenly required to do so. In order to address this kind of case, two types of operating periods can be employed in the optimisation framework to distinguish between short and long-term intervals. For instance, units might be allowed to freely modify their operation between seasonal periods (e.g. months), whereas they would have some limitations for inter-seasonal ones (e.g. hours of the day) so that the equipment is prevented from drastic load variations. Thus, Eq. (4) can be employed to indicate that if a unit is operating on an inter-seasonal period (t,θ), then, on the next one (i.e., $t,\theta+1$), it can either be switched off or its load could only be reduced up to a minimum turndown value. However, if it is not operating during (t,θ), it would not be able to be started on the $(t,\theta+1)$ period.

On the other hand, it is also possible to include (3) and (4) in the optimisation of retrofit or grassroots design cases. In these types of problems new equipment might be installed and all the operating scenarios are taken into account to determine a final configuration that can cope with all of them. Consequently, the most convenient (e.g. cost-effective) redundancy due to equipment maintenance is established during the solving procedure without the need of pre-defining any element of the final design. Although, failure conditions have not been considered yet, the following section will describe how to account for these issues into the design and operation of utility systems:

$$\sum_t z_{n,t,\theta}^{op} \leqslant (T - Num_n^{maint}), \quad t = 1, 2, \ldots, T$$

$$(3)$$

$$Output_{n,t,\theta+1} \geqslant TurnDown_n^{max} \cdot Output_{n,t,\theta}$$

$$(4)$$

Redundancy Due to Failure Situations

Reliability and availability issues in the design and operation of utility systems can only be addressed within an integrated methodology when normal, maintenance, and failure conditions are taken into account simultaneously. So far it has been explained how maintenance situations (i.e., scheduling) can be automatically defined while (at the same time) the required redundancy to deal with them is also established by the optimisation. Accordingly, the next step is to incorporate failure scenarios into the time horizon and to employ the suggested method to obtain a final design with enough redundancy to cope with all circumstances and satisfy the specified demands, while minimising/maximising the aim of the optimisation task (e.g. reducing costs).

As mentioned previously, although the number of possible failure modes for a typical utility plant is overwhelming, the majority of them are highly improbable and can be neglected. For instance, it might be evident that for a system comprised of 10 units, more than one of them failing (independently) at the same time would be very unlikely and, thus, the failure modes could be cut from $2^{10}=1024$ to just 10 (for a certain time interval). Even more cuts could be done if not all the units are (functionally) of the same type and/or if some units of the same type are also of the same size. With the suggested approach, users have the possibility of evaluating several situations for a specific application and they can also compare the results from different assumptions. In the case of pure operational problems, once an initial optimisation has yielded the most convenient operational strategy for all periods considering only equipment maintenance (i.e., scheduling), it is then necessary to establish (e.g. with a reliability analysis like inEbeling, 1997) which are the relevant failure circumstances to be incorporated and their corresponding probabilities. These conditions are then set on additional operating scenarios indicating the specific units assumed to breakdown (i.e., by assigning a value of zero to the corresponding operational binary variables). The relative length of the failure periods would depend on their probabilities of incidence. Also note that these short failure periods should not be included in the scheduling of the units (i.e., Eq. (3)). In this way it can be investigated what are most convenient operating strategies for an existing plant during normal, maintenance, and failure situations, and/or check whether the system can cope with all of them whilst fulfilling the utility requirements.

On the other hand, in retrofit and grassroots design problems the final configuration of the utility system, including redundant elements, is not known beforehand, and failure situations have to be incorporated with a different approach than in operational cases. The first step is to identify within the superstructure (of design options) the pieces of equipment performing a common function (e.g. boilers producing high-pressure steam). Units performing more than one function (e.g. gas turbines produce power and can raise steam in a heat recovery unit) would belong to several functional types. Then, it is possible to apply Eq. (5)to indicate that if several units of the same (functional) type are selected and sized by the optimisation, they would be assigned a series of numbers in descending order according to their sizes (i.e., maximum outputs). Note that this constraint is not pre-specifying the elements (e.g. equipment sizes) of the plant at all, and it is simply ensuring that there will be a mathematical way of identifying whether a unit (if selected) would be greater than another one of the same type (if both are selected).

The next step is to specify the pertinent failure scenarios to be included in the optimisation given that it would not be viable to enumerate all of them. Also, defining such situations cannot be done explicitly for retrofit or grassroots design cases since the final configuration is still unknown. However, in this kind of problems the major concern is to establish the appropriate redundant elements of the utility system rather than finding the best operating conditions for a particular scenario. Once a design has been established, any other conditions can be investigated with an additional optimisation (i.e., an operational task). In view of such priority, the suggested strategy is to incorporate in the design task only certain (critical) failure situations defined in such a way that the redundancy provided by the optimisation will be enough not only to deal with them, but also with any other operating scenarios (normal, maintenance, and/or failure ones) that might also happen:

$$Output_n^D \geqslant Output_{n+1}^D \tag{5}$$

$$Output_{A1,t\text{-fail},\theta\text{-fail}} = 0 \tag{6}$$

$$Output_{A2,t\text{-fail},\theta\text{-fail}} = 0 \tag{7}$$

In order to identify these critical failure situations there are three factors to assess: the number of units (of the same functional type) that are assumed to be down, the relative sizes of such units, and the periods when the failures are supposed to occur. Accordingly, critical failure situations (i.e., the ones that will require the largest degree of redundancy) can be regarded as those for which:

- The number of shutdown units of the same functional type is the maximum practical one. Users should define this amount according to the particular application, but, generally, assuming a maximum of two or three units down at the same time (e.g. two failing while one is on maintenance) is enough for most problems, unless the number of potential units of the same type or their possibility of failure is very large. In any case, it is always possible to investigate different numbers and compare the results.

- The failing units are the largest ones of each functional type. Although the final design is not known yet, by virtue of Eq. (5) it is possible to specify that the largest units (e.g. the first two largest) of each type (if selected) should be down for a given failure situation. For example, (6) and (7) could be employed to impose that the largest and second largest units (of unknown sizes) of a certain type must be switched-off (if selected) during the failure scenario (t-fail,θ-fail). Similarly, as with operational tasks, the relative duration of a failure scenario will depend on its probability of incidence, which can be obtained by conventional means (e.g. reliability methods such as the ones presented byEbeling, 1997). Also note that the design is not being pre-defined in any way, because the optimisation might still decide not to select any units of that type at all.

- The failure takes place during time periods when the requirements of one or more utilities are the highest. Mathematically speaking, in order to guarantee that all the critical situations (i.e., for every type of equipment) are considered in the optimisation, it is necessary to include, for each normal period, a set of failures in which the largest (say) three units (of each type) are assumed to be down. Although this exhaustive approach might be feasible for some s mple problems, it would not be practical for complex utility systems. Nevertheless, it is very likely that the redundancy required to cope with a certain number of units down, will be maximum if the failure occurs when the demands of one or/and some utilities are the largest. Moreover, in most cases it would be evident which utility (or sum of utilities) is critical for a given functional type of equipment (e.g. overall steam consumption for VHP steam generators) and, thus, it would be straightforward to define which and how many normal periods should be tested for the failure of a given type of equipment. In an extreme case, it is always possible to run the problem several times with different assumptions and compare the results.

As can be noted, none of the previous factors is implying the specification of any design elements (including redundant ones). Therefore, with the proposed strategy users can introduce critical failure scenarios for a given (retrofit or grassroots) problem, without manipulating the final configuration. In this way, the optimisation will decide the number, types and sizes of any redundant components together with the rest of the units of the plant. In other words, once normal and failure scenarios have been introduced into the time horizon of the problem, the optimisation will determine simultaneously the design of the whole utility plant (including redundant elements), the scheduling for every piece of equipment, and the operating strategy for the corresponding normal, maintenance, and failure conditions so that the specified demands will be always fulfilled, while minimising costs. The detailed mathematical expression for optimisation frameworks can be found from Aguilar et al., 2007a and Aguilar et al., 2007b, in which constraints explained in this Section need to be complemented for considering redundancy.

CASE STUDY

This case is based on a project to build a large grassroot complex. Although the site is being designed from scratch, the various processing plants together with the pressures of the headers to distribute steam have already been decided (the major site conditions for this case are presented on Table 1). Similarly, it has been possible to estimate, for the base operation of the complex, a set of approximate demands, which are expected to be also quite large (see Table 2). Moreover, since the utility system should be, in principle, self-sufficient, it must be able to fully meet a base demand of 540 MW of electricity without importing (or exporting) power. It also must supply 1000 MW of steam at five different levels, and 2000 MW of cooling water. Note that the steam to be consumed and/or generated by processes is at near saturated conditions (except for the VHP header).

Table 1: Site data for the case study

Site conditions		
Total working hours	hrs/yr	8600
Annualisation factor	–	0.10
Ambient temperature	°C	25

Altitude	m	0
Relative humidity	%	60
Fuel oil #2 LHV	kJ/kg	45 000
Natural gas LHV	kJ/kg	50 244
Electricity price	$/kWh	0.05
Fuel oil #2 prices	$/kg	0.22
Natural gas prices	$/kg	0.33
Raw water prices	$/ton	0.18

Table 2: Demands data for the case study

Requirements for the utility system		
Electricity demands	MW	540
VHP steam demands	kg/s	0.0
VHP steam generation	kg/s	18.5
HP steam demands	kg/s	108.2
HP steam generation	kg/s	35.6
MP steam demands	kg/s	213.9
MP steam generation	kg/s	11.0
LP steam demands	kg/s	107.6
LP steam generation	kg/s	0.0
Net steam demands	MW	1000
Condensate return	%	70
Process CW demands	MW	2000

The objective of a grassroots design problem such as this one is to establish all the equipment to be installed (i.e., types and number of units, their connections, and sizes) together with the operating conditions of the whole utility plant that can satisfy the specified demands, while minimising a certain optimisation aim. The corresponding objective function for this analysis is total cost, comprised of the overall operating expenses plus the required capital multiplied by an annualisation factor of 0.10 (i.e., the annualised total cost is the summation of annual operating cost and 10% of the investment.).

In addition, a realistic design should also take into account that the units of the utility system need to be switched off to receive maintenance and might also experience unexpected failures. However, as discussed in the introduction, reliability and availability concerns are barely accounted for in conventional methods addressing the design of energy systems. Conversely, within the suggested approach it is possible to incorporate these issues given that the design and operational variables are optimised simultaneously considering various scenarios. Thus, for this case study several examples with different assumptions regarding equipment reliability and availability will be compared in order to demonstrate the importance of these issues.

Before describing the different examples for this case, it is first necessary to determine the steam conditions of the headers that are employed to generate power inside steam turbines. Note that, with the present methodology, it is possible to define different properties for the steam supplied to processes and for the steam used in power production. In fact, while process steam is normally required at near-saturated conditions, for the second case the steam must have some degree of superheating to avoid wetness in turbine discharges. Therefore, the procedure explained in Aguilar (2005) was applied for this case, assuming common pressures, and the resulting steam properties are summarised on Table 3. In this way, all the potential back-pressure turbines (if selected) would discharge steam at or slightly higher conditions than the headers, and the exhaust from condensing units would not result in dryness below 90%.

Table 3: Steam conditions for the case study

Steam conditions			
VHP pressure		bara	104.4
HP pressure		bara	42.4
MP pressure		bara	11.4
LP pressure		bara	4.5
Deaerator pressure		bara	1.1
Condenser pressure		bara	0.98
Condensate temp.		°C	90
		Processes	Turbines
VHP temperature	°C	540	540
HP temperature	°C	254	370
MP temperature	°C	186	245
LP temperature	°C	150	160

Design without Elements of Redundancy

As discussed previously, an initial design assuming that all the pieces of equipment within the utility system are always available will be employed as base case for comparative purposes. Hence, a single operating scenario (with the expected set of demands) was defined for such an ideal situation, considering that no elements of redundancy are required. The corresponding optimisation model consisted of 760 equations, 581 variables, of which 52 were binary, and it took 764 major iterations to arrive at a solution with 0.005% of relative gap (GAMS® 2.0, Cplex solver; Brooke et al., 1998) in 0.29 s of CPU time (Pentium IV processor at 3.0 GHz).

As can be observed from the flowsheet on Fig. 6, the optimal design for this example features two gas turbines of 200.0 MW with supplementary-fired (SF) HRSG (each one of 171.3 kg/s) plus a complementary 34.0 MW GT with an unfired HRSG of 14.7 kg/s. Note that 200.0 MW has been set as the maximum size for gas turbines, and that these units are slightly de-rated due to the high ambient temperature. In this way, given the relative large amount of power required with respect to the heating demands, only GT+HRSGunits are employed to raise steam (injected to the VHP header), which is then expanded through six back-pressure turbo-generators before being delivered to external consumers. Therefore, not only the letdown flows through the valves are very small, but also the use of condensing turbines and steam venting have been avoided so that the system can operate in a full cogeneration mode (i.e., despite the installation of a cooling water system to satisfy process demands).

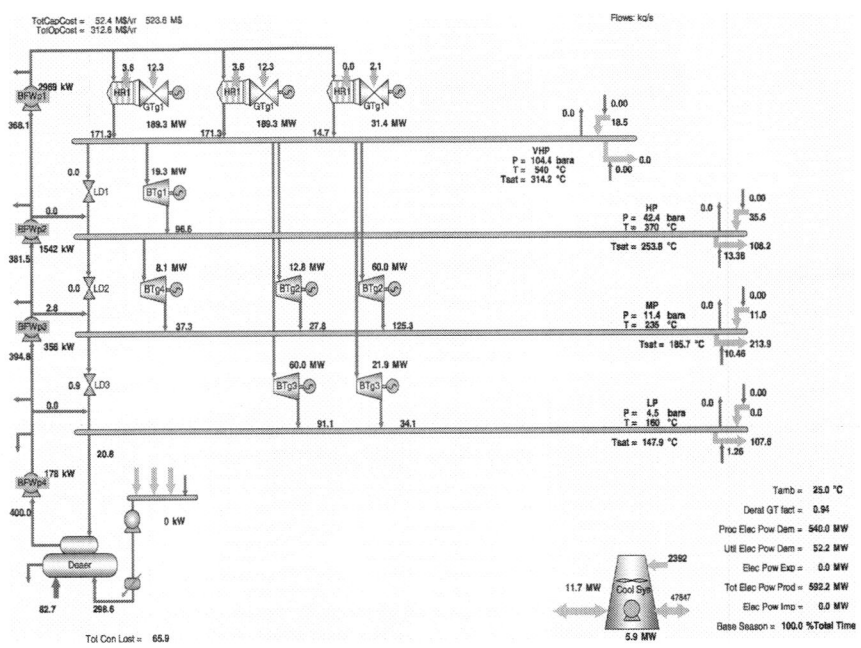

Figure 6: Flowsheet for case study without considering maintenance or failure redundancy (base operation).

In this sense it is worth mentioning that if the third GT is eliminated, then a boiler and a condensing turbine would be needed, but such a non-cogeneration scheme implied large cost penalties. It is also interesting to note that the optimisation determined to use two GT of the largest size possible because they are more efficient and cheaper (per kW) than smaller units. The required investment for this design without accounting for reliability and availability issues is $523.6 millions and its overall operating cost corresponds to $312.6 millions/yr.

Design Considering Maintenance Redundancy

For the second example of this case study the maintenance of boilers and gas turbines has been considered as an initial approach to address reliability and availability issues during the design of the plant. For this purpose the time horizon of the problem (i.e., one year) has been divided into three periods with the same set of demands and site conditions. While the first scenario corresponds to the normal operation of the plant (i.e., all the units available), in the second one the largest GT (if selected) has been forced to shut down and, during the third one, the largest boiler (if selected) must be switched off to receive preventive maintenance. Note that, since demands variations are not available, a detailed scheduling for all units would not be practical at this point, and thus the relative length of the three periods has been initially estimated as 70-20-10% (an scheduling example has been presented in Aguilar et al., 2007b). The corresponding optimisation task consisted of 2121 equations, 1289 variables, of which 150 were binary, and it took 582 863 major iterations to arrive at a solution with 0.01% of relative gap (GAMS® 2.0, Cplex solver; Brooke et al., 1998) in 83.8 s of CPU time (Pentium IV processor at 3.0 GHz). The optimal flowsheets for the three scenarios of this example are presented on Fig. 7. The final design features four gas turbines of 112.3 MW with SF-HRSG units of 95.6 kg/s together with two boilers of 100.0 kg/s to supply steam at VHP level. It is interesting to observe how the optimisation determined (i.e., without any size or number pre-specification[1]) to equally distribute the sizes for the units of the same type subject to maintenance, so that the system is not relying more on any specific item. Also, as in the previous case, most of the steam is expanded inside back-pressure turbines (i.e., eight major units plus two small ones) to avoid large flows through letdown valves. However, now it is necessary to exploit the part-load flexibility of the equipment to cope with maintenance situations and operate the system in a cogeneration scheme for most of the time. For instance, while only one HRSG burns supplementary fuel during the base scenario, two recovery units run in this mode to compensate for the shortage of steam when one of the boilers is not working. Also, while boilers and steam turbines normally operate at part-load, they tend to run at full capacity during the maintenance periods.

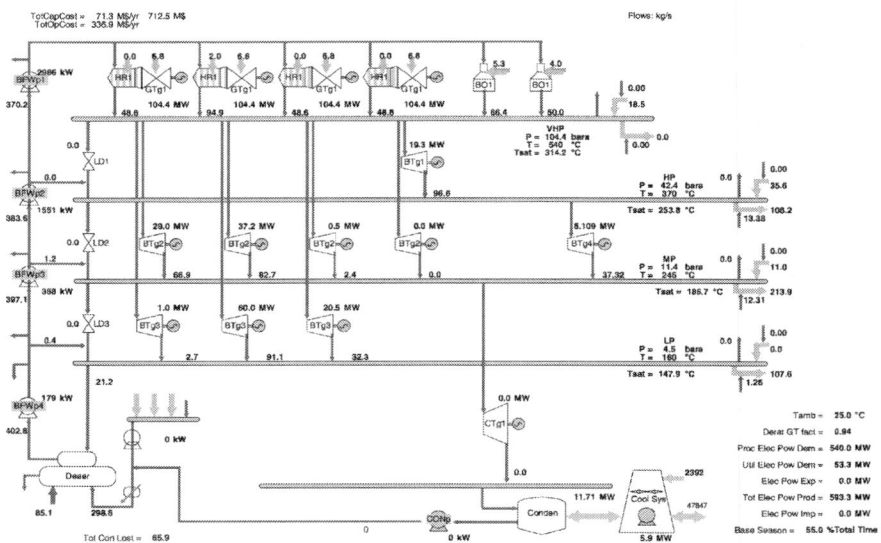

(a)

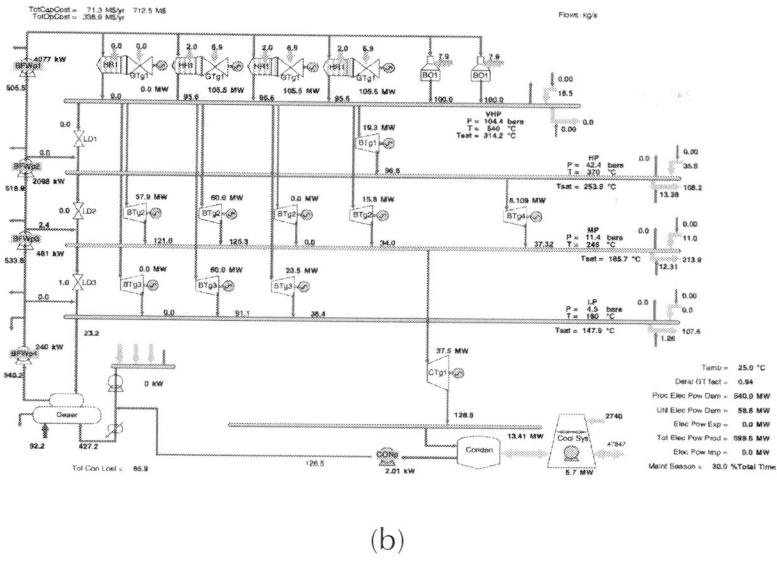

(b)

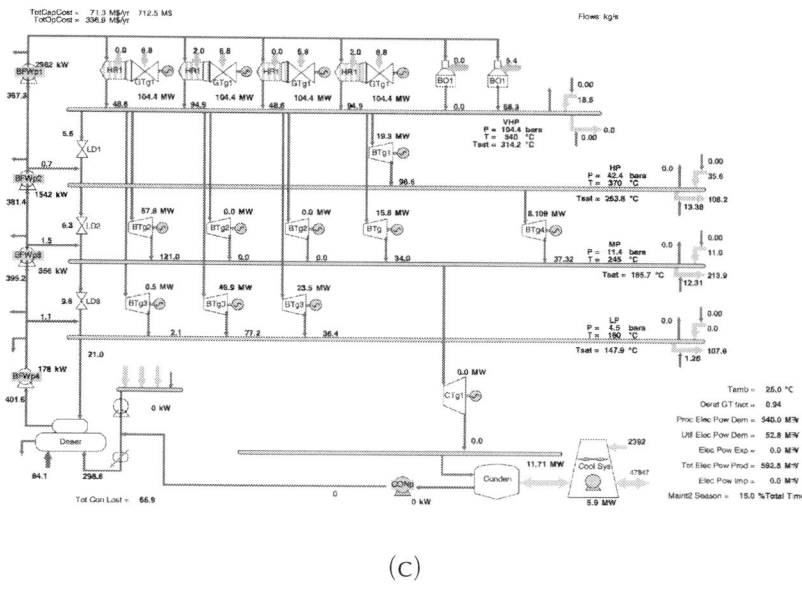

(c)

Figure 7: Flowsheets for the case study considering maintenance redundancy: (a) base operation, (b) one GT shutdown, and (c) one boiler shutdown.

Additionally, when one of the GT is down for maintenance, it is necessary to increase the load of the steam turbines to produce more power. Consequently, even though the other three gas turbines are operating at full load and their heat recovery units are burning supplementary fuel, the two boilers must also run at full capacity. Hence, for this scenario there is an excess of steam beyond process requirements that, after producing power inside back-pressure turbines, need to be expanded through a condensing one (i.e., a non-cogeneration scheme). On the other hand, note that the relative length for the time periods was adjusted after a first optimisation according to the number of gas turbines and boilers being selected (i.e., to 55-30-15% assuming that each of these units is down once a year). Finally, regarding the economic performance of the plant, the required investment for this example is $712.5 millions and its overall operating cost corresponds to $336.9 millions/yr. Therefore, accounting for equipment maintenance represented an additional capital of $188.9 millions (36.1%) and an increase in operational costs of $24.3 millions/yr (7.8%) in comparison to the base case.

Design Considering Maintenance and Failure Redundancy

In this example availability and reliability issues are addressed during the design of the utility plant by considering not only the maintenance for major pieces of equipment units, but also situations of unexpected failure involving these units. As explained previously, from all the probable failure modes of a utility system only a few scenarios are likely to occur in practice. Moreover, for design problems the main objective is to determine the configuration of the plant that will be able to cope with a certain critical circumstances rather than investigating how to operate the system for an extensive number of specific conditions (i.e., this can be done afterwards with an operational optimisation run).

Consequently, for this example, it has been assumed that the probability of the simultaneous failure of two or more steam-producing units can be neglected, but it is possible that one of these units trips while another is receiving maintenance. Thus, the time horizon of the problem (i.e., one year) has been divided into four periods corresponding to the normal operation of the system, the scheduled shutdown of a GT, the failure of a GT when a boiler is down for maintenance, and the failure of a GT when another one is not working. Note that, as explained previously, in order to account for the critical situations, the units forced to switch off in all these scenarios are the largest of each type.[2] This optimisation task consisted of 2816 equations, 1643 variables, of which 201 were binary, and it took 425 584 major iterations to arrive at a solution with 0.05% of relative gap (GAMS® 2.0, Cplex solver; Brocke et al., 1998) in 84.2 s of CPU time (Pentium IV processor at 3.0 GHz).

The optimal flowsheet for the four scenarios of this example are presented on Fig. 8. The final design includes four gas turbines of 109.3 MW with SF-HRSG units of 93.1 kg/s and five boilers (four of 100.0 kg/s and one of 54.4 kg/s) supplying steam to the VHP header. As in the previous case, the optimisation decided to distribute equipment sizes evenly (except for the last boiler) between the units subject to maintenance and/or failure in order to avoid relying on any particular one. Also, most of the steam is expanded inside nine back-pressure turbines to reduce the flows through letdown valves. However, due to the large operating variations caused by the failure situations, the part-load flexibility of the equipment is not enough to prevent some steam to be condensed in all periods (i.e., in a non-cogeneration scheme). For instance, while only the smallest condensing turbine (1.4 MW) is normally operating, one of the large ones (41.6 MW) is running for the second and third periods, and the three of them must expand steam during the fourth scenario. Moreover, the loads for back-pressure turbines and boilers, along with the number of HRSG running in SF mode, increases as it is required to compensate for more steam and power generation.

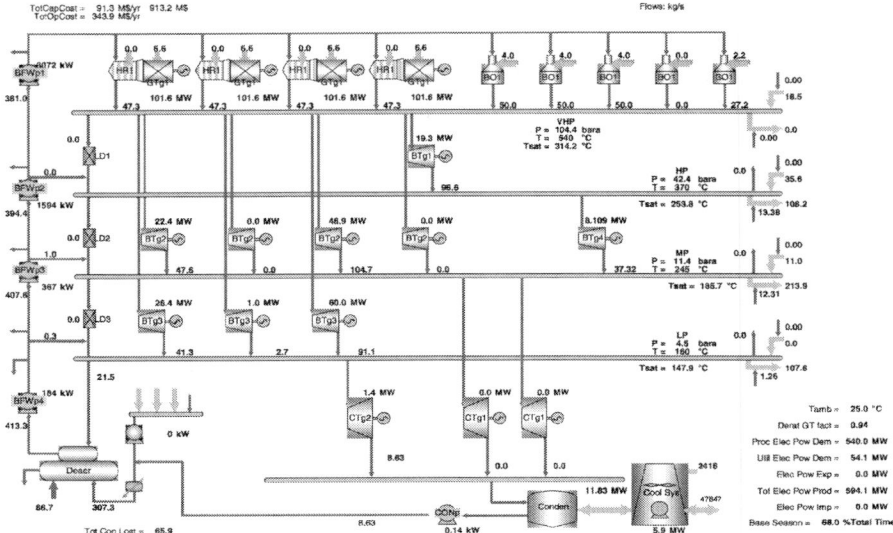

(a)

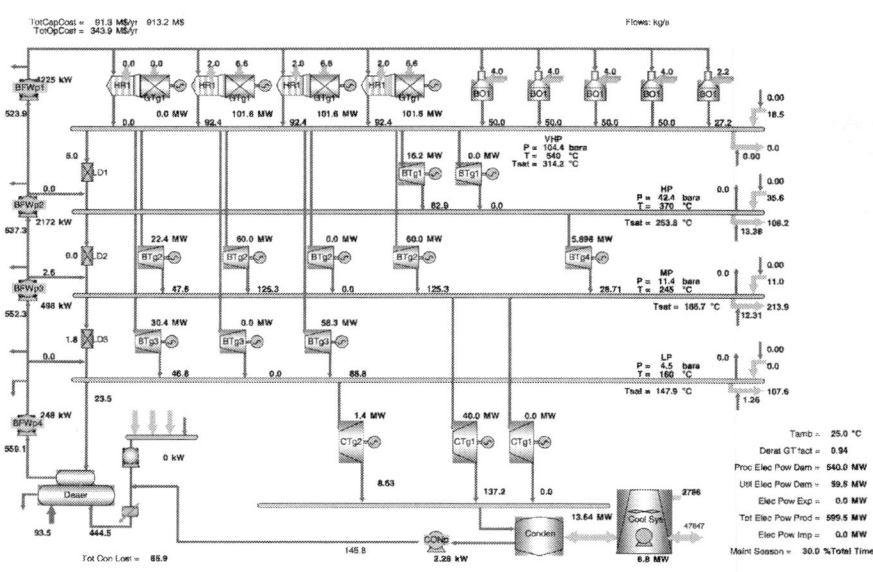

(b)

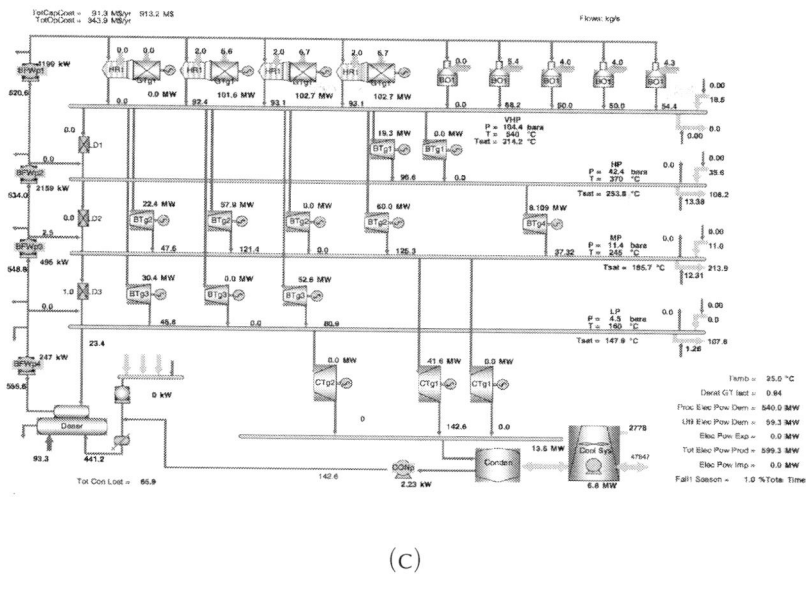

(c)

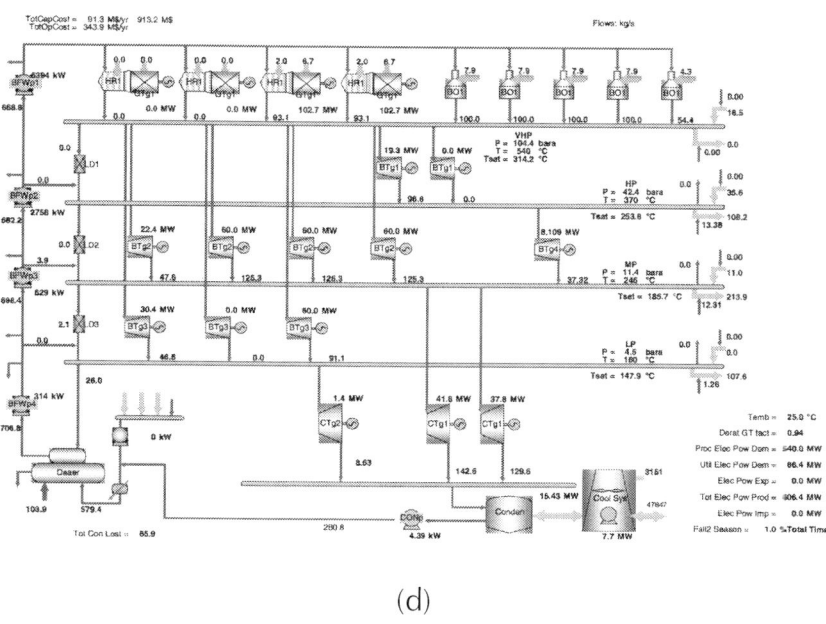

(d)

Figure 8: Flowsheets for the case study considering maintenance and failure redundancy (no power import allowed): (a) base operation, (b) one GT shutdown, (c) one boiler and one GT shutdown, and (d) two GT shutdown.

On the other hand, in this example, the relative length for the time periods was again adjusted after a first optimisation (according to the number of gas turbines and boilers being selected) to 68-30-1-1% where the small percentages were defined just to make sure that the plant can cope with (emergency) failure conditions. Finally, regarding the economic performance of the plant, the required investment for this case is $913.2 millions and its overall operating cost corresponds to $343.9 millions/yr. Therefore, accounting for equipment maintenance and failure represented an additional capital of $389.6 millions (74.4%) and an increase in operational costs of $31.3 millions/yr (10.0%) taking as reference the base case.

Additionally, the effect of allowing 50 MW of emergency power import was investigated for the scenario when two gas turbines are down (i.e., one for maintenance and the other for failure). With this supplementary element of redundancy the utility system has to cope with less pronounced variations and now it was possible to install less pieces of equipment and the part-load flexibility of the plant has avoided condensing steam for most of the time. Moreover, the required capital and the corresponding operating cost are lower than in the previous case (i.e., $820.3 millions and $344.1 millions/yr), but this needs to be further assessed against any applicable connection costs to the electric grid (Fig. 9).

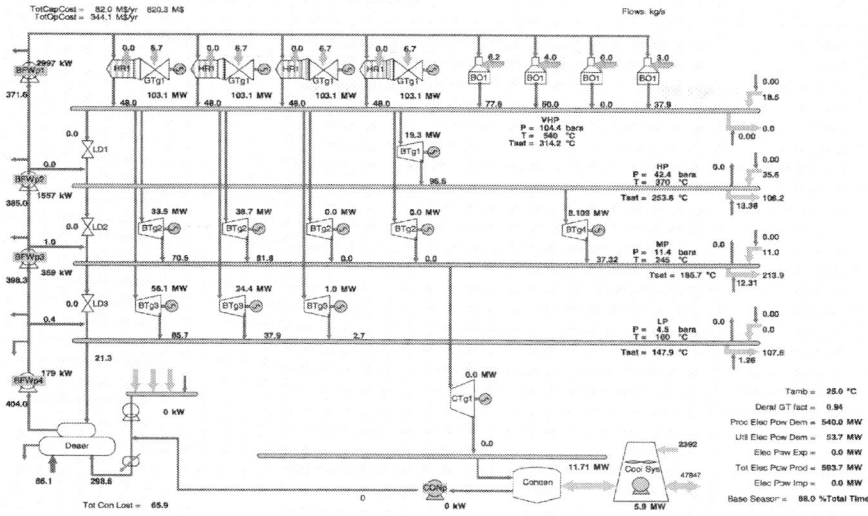

(a)

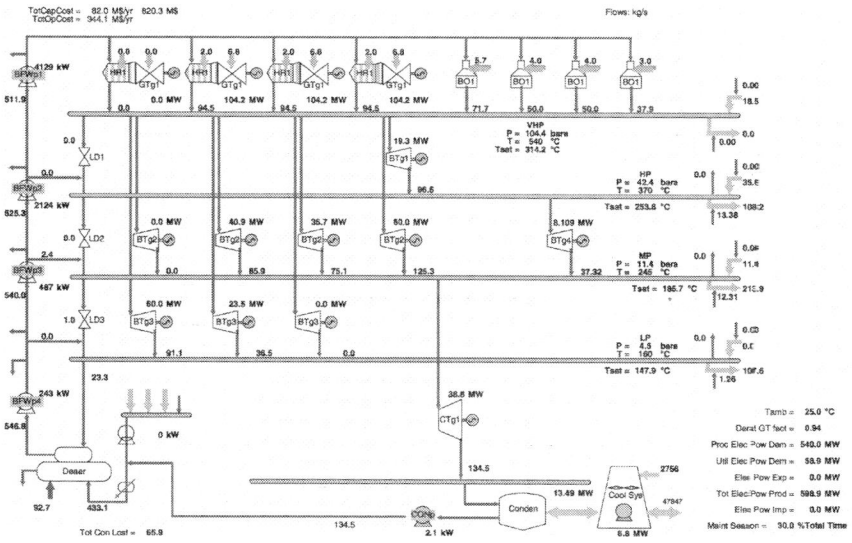

(b)

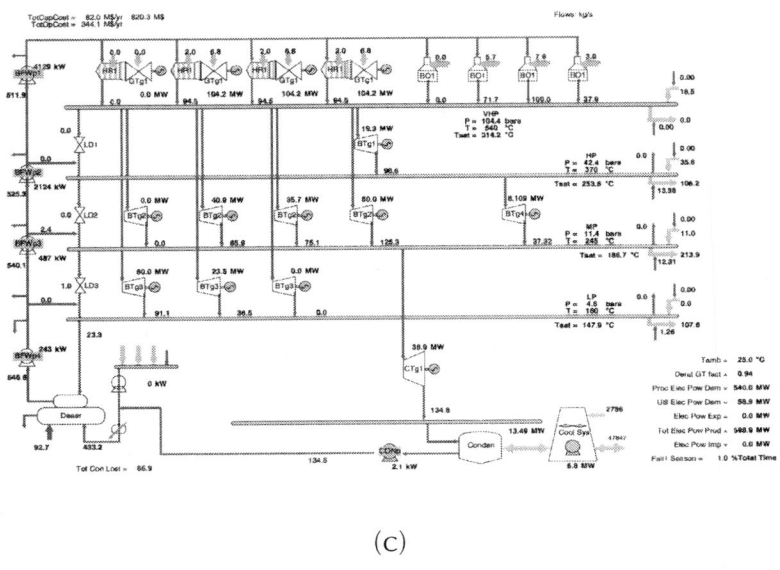

(c)

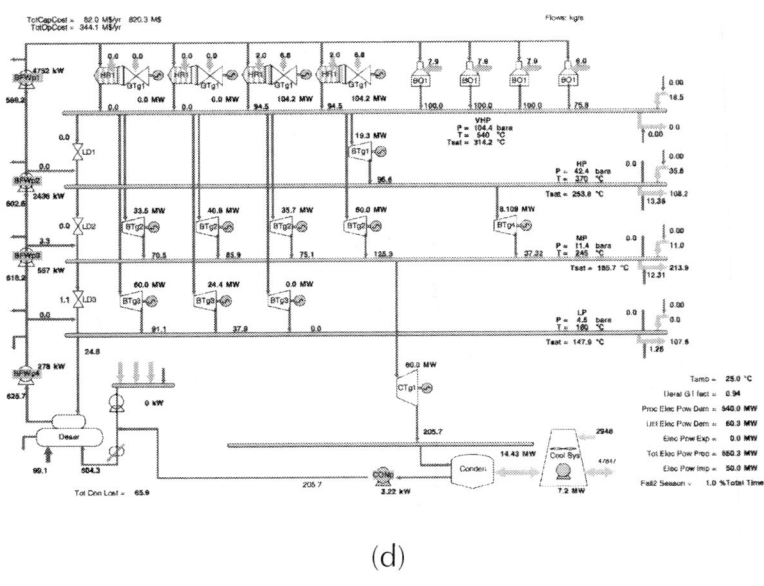

(d)

Figure 9: Flowsheets for the case study considering maintenance and failure redundancy (50 MW imported for failure cases): (a) base operation, (b) one GT shutdown, (c) one boiler and one GT shutdown, and (d) two GT shutdown.

CONCLUSIONS

In order to provide realistic solutions, availability and reliability issues must be taken into account when addressing the design and operation of utility plants. However, most of the previous approaches have neglected these concerns or used rough rules of thumbs, and just a few pieces of research have employed programming techniques, but with limited application to practical cases. Conversely, the present work provides an integrated methodology that includes maintenance and failure situations during the optimisation of the design and operating variables of industrial utility systems (i.e., for grassroots, retrofit, and operational problems). In the case of an existing plant, the procedure can establish the operating conditions together with the maintenance scheduling of every piece of equipment. Also, it is possible to investigate whether the system can cope with certain failure scenarios and what would be the best operating strategy for all the circumstances being evaluated. Regarding retrofit and grassroots design problems, the proposed method allows introducing critical failure conditions without pre-specifying any options in the final configuration.

Therefore, the most convenient elements of redundancy are obtained through optimisation as part of the whole plant design. In these cases also, the scheduling and the operating conditions for all the scenarios are automatically established so that the expected demands are always fully satisfied while minimising, for example, overall capital and operating costs. In this way, the suggested approach not only offers a significant potential of reducing operating costs, but also of achieving large investment savings since it ensures the strategic allocation of any available capital.

REFERENCES

1.   Aguilar, O., 2005. Design and optimisation of flexible utility systems. Ph.D. Thesis, The University of Manchester, Manchester, UK.

2. Aguilar, O., Perry, S., Kim, J., Smith, R., 2007a. Design and optimisation of flexible utility systems subject to variable conditions---part 1: modelling framework. Chemical Engineering Research and Design 85 (A8), 1136--1148.

3. Aguilar, O., Perry, S., Kim, J., Smith, R., 2007b. Design and optimisation of flexible utility systems subject to variable conditions---part 2: methodology and applications. Chemical Engineering Research and Design 85 (A8), 1149--1168.

4. Ashok, S., Banerjee, R., 2003. Optimal operation of industrial cogeneration for load management. IEEE Trans. Power Syst. 18, 115--126.

5. Brooke, A., Kendrik D., Meeraus A., Raman, R., 1998. GAMS---A User's Guide. GAMS Development Corporation.

6. Bruno, J.C., Fernandez, F., Castells, F., Grossmann, I.E., 1998. A rigorous MINLP model for the optimal synthesis and operation of utility plants. Chemical Engineering Research and Design 76, 246--258.

7. Del Nogal, F., Kim, J., Perry, S., Smith, R., 2005. Systematic driver and power plant selection for power-demanding industrial processes. In: A.I.Ch.E. Spring Meeting, Atlanta, US. Ebeling, C., 1997. An Introduction to Reliability and Maintainability Engineering. McGraw-Hill, New York.

8. Frangopoulos, C.A., Dimopoulos, G.G., 2004. Effect of reliability considerations on the optimal synthesis, design and operation of a cogeneration system. Energy 29, 309--329. Govil, A.K., 1983. Reliability Engineering. McGraw-Hill, New Delhi.

9. Hui, C., Natori, Y., 1996. An industrial application using mixed-integer programming technique: a multi-period utility system model. Computers and Chemical Engineering 20, S1577--S1582.

10. Iyer, R.R., Grossmann, I.E., 1997. Optimal multiperiod operational planning for utility systems. Computers and Chemical Engineering 21, 787--800.

11. Maia, L.O.A., Qassim, R.Y., 1997. Synthesis of utility systems with variable demands using simulated annealing. Computers and Chemical Engineering 21, 947--950.

12. Maréchal, F., Kalitventzeff, B., 2003. Targeting the integration of multi-period utility systems for site scale process integration. Applied Thermal Engineering 23, 1763--1784.

13. Nath, R., Holliday, J.F., 1985. Optimizing a process plant utility system. Mechanical Engineering, February, 44-50.

14. Olsommer, B., Favrat, D., Spakovsky, M.R., 1999a. An approach for the time-dependent thermoeconomic modeling and optimization of energy system synthesis, and operation--part 1. International Journal of Applied Thermodynamics, 2 (3), 97--113.

15. Olsommer B., Favrat D., Spakovsky M.R., 1999b. An approach for the time-dependent thermoeconomic modeling and optimization of energy system synthesis, and operation--part 2. International Journal of Applied Thermodynamics, 2 (4), 177--186.

16. Papoulias, S.A., Grossman, I.E., 1983. A structural optimization approach in process synthesis---I. Utility systems. Computers and Chemical Engineering 7, 695--706.

17. Varbanov, P.S., Doyle, S., Smith, R., 2004. Modelling and optimisation of utility systems. Chemical Engineering Research and Design 82 (A5), 561--578.

18. Varbanov, P.S., Perry, S., Klemes, J., Smith, R., 2005. Synthesis of industrial utility systems: cost-effective de-carbonisation. Applied Thermal Engineering 25, 985--1001.

Citations

CHAPTER 1

Peng Li, Hideki Abe, and Jinhua Ye, "Band-Gap Engineering of NaNbO3 for Photocatalytic H2 Evolution with Visible Light," International Journal of Photoenergy, vol. 2014, Article ID 380421, 6 pages, 2014. doi:10.1155/2014/380421.

CHAPTER 2

Alaneme, K., Bamike, B. and Omlenyi, G. (2014) Design and Performance Evaluation of a Sustained Load Dual Grip Creep Testing

Machine. Journal of Minerals and Materials Characterization and Engineering, 2, 531-538. doi:10.4236/jmmce.2014.26054.

CHAPTER 3

N. Othman and S. K. Kamarudin, "Radiotracer Technology in Mixing Processes for Industrial Applications," The Scientific World Journal, vol. 2014, Article ID 768604, 15 pages, 2014. doi:10.1155/2014/768604.

CHAPTER 4

Hyunsu Kim, Jin-Seo Noh, Jong Wook Roh, Dong Won Chun, Sungman Kim, Sang Hyun Jung, Ho Kwan Kang, Won Yong Jeong, and Wooyoung Lee, Perpendicular Magnetic Anisotropy in FePt Patterned Media Employing a CrV Seed Layer, doi:10.1007/s11671-010-9755-2.

CHAPTER 5

Michael C. Georgiadis, Gordian Schilling, Guillermo E. Rotstein, Sandro Macchietto, A general mathematical programming approach for process plant layout, Computers & Chemical Engineering, Volume 23, Issue 7, 1 July 1999, Pages 823-840, ISSN 0098-1354, http://dx.doi.org/10.1016/S0098-1354(99)00005-8.

CHAPTER 6

Sebastian Terrazas-Moreno, Ignacio E. Grossmann, John M. Wassick, Scott J. Bury, Naoko Akiya, An efficient method for optimal design of large-scale integrated chemical production sites with endogenous uncertainty, Computers & Chemical Engineering, Volume 37, 10 February 2012, Pages 89-103, ISSN 0098-1354, http://dx.doi.org/10.1016/j.compchemeng.2011.10.005.

CHAPTER 7

E. Godoy, S.J. Benz, N.J. Scenna, An optimization model for evaluating the economic impact of availability and maintenance notions during the synthesis and design of a power plant, Computers & Chemical Engineering, Volume 75, 6 April 2015, Pages 135-154, ISSN 0098-1354, http://dx.doi.org/10.1016/j.compchemeng.2015.01.020.

CHAPTER 8

Oscar Aguilar, Jin-Kuk Kim, Simon Perry, Robin Smith, Availability and reliability considerations in the design and optimisation of flexible utility systems, Chemical Engineering Science, Volume 63, Issue 14, July 2008, Pages 3569-3584, ISSN 0009-2509, http://dx.doi.org/10.1016/j.ces.2008.04.010.

Index